The Social Sciences Go to Washington

The Social Sciences Go to Washington

The Politics of Knowledge in the Postmodern Age

EDITED BY

HAMILTON CRAVENS

RUTGERS UNIVERSITY PRESS
New Brunswick, New Jersey, and London

Library of Congress Cataloging-in-Publication Data

The social sciences go to Washington : the politics of knowledge in the postmodern age / edited by Hamilton Cravens.

p. cm.

Includes bibliographical references and index.

ISBN 0–8135–3340–6 (hardcover: alk. paper)—ISBN 0–8135–3341–4 (pbk.: alk. paper)

1. Political planning—United States. 2. Policy scientists—United States. 3. Social scientists in government—United States. I. Cravens, Hamilton.

JK468.P64S63 2004
320'.6'0973—dc21

2003005991

British Cataloging-in-Publication information is available from the British Library.

Manufactured in the United States of America

Contents

Preface and Acknowledgments

THIS VOLUME BEGAN as an idea over lunch with my colleague, Alan I Marcus, in the only Italian restaurant in all of beautiful downtown Ames, Iowa. We wanted another volume of essays that reflected the kinds of work going on in our department's program in History of Technology and Science, and I suggested the topic of the social sciences go to Washington in the several decades after World War II; after some discussion, we came to mean what were the influences of the social sciences upon social policies and their making, what social science ideas became institutionalized in policy, and how and why. We were not so interested in the ideas of expertise and professionalism inherent in the social sciences as knowledge systems, as that had been done and done again; we wanted something less stale, less well known, less familiar. We identified the areas of policy to be covered, fully realizing that we could not cover everything, or anything in adequate detail, in such a volume as this. Then we wrote potential contributors, all of whom responded generously and with great intelligence and energy. We tried out our ideas at the tenth anniversary conference of the *Journal of Policy History* in 1999 in St. Louis. What began as trial runs soon became polished chapters. Though Alan Marcus left the project to pursue other interests, the contributors are grateful to him for his efforts in identifying a suitable publisher.

I would like to thank my co-contributors for their amiable cooperation with all the tasks at hand attendant to publication. I would also like to express my appreciation to the two anonymous referees for the Rutgers University Press, who wrote trenchant, useful, and constructive reports that were an important guide to me and the other contributors, as well as to Audra Wolfe, science editor, and her colleagues, for a marvelous experience in publishing. A generous grant in aid of publication from the Office of the Vice Provost for Research and Advanced Studies of Iowa State University is gratefully acknowledged. Thanks also to Andrejs Plankans, chair of the History Department at Iowa State University, for financial assistance with preparing the index and with other overall editorial tasks. Robert Burchfield did a fine job of copy-editing, and Margie Towery deserves credit for indexing this book. Two of the essays, those by William Graebner and myself, have been previously published.

HAMILTON CRAVENS gratefully acknowledges the permission of Jack Salzman, editor of *Prospects*, for permission to republish "American Social Science and the Invention of Affirmative Action, 1920s–1970s," which first appeared in *Prospects* 26 (2001): 361–389. William Graebner wishes to acknowledge the permission of the Pennsylvania State University Press to republish "The End of Liberalism: Narrating Welfare's Decline, from the Moynihan *Report* (1965) to the Personal Responsibility and Work Opportunity Act (1996)," originally published in *Journal of Policy History* 14, no. 3 (2002): 171–190.

The Social Sciences Go to Washington

Introduction

The Social Sciences, the Federal Government, and the Age of Postmodernism

HAMILTON CRAVENS

AH, SCIENCE, that wondrous human activity whose practitioners promise us more and more fabulous and exciting views of nature and society. As seductive have been and are the tangible products and results of the application of scientific knowledge to particular problems of nature and of society. What a sublime legacy of the last two centuries, so educated and cultured Americans could easily believe. With the rise of professionalized, university-based science in the last quarter of the nineteenth and the first third of the twentieth centuries, Americans could look with pride upon their national culture and its ability to keep up with and to surpass its sister European national cultures in the arts and the sciences. Institutions, whether of government, of the academy, of the private business world, of eleemosynary intent, or of culture, now functioned as parts of a national system of commerce and trade, government and intergroup relations, artistic and scientific creativity, the application of invention and imagination to all manner of human economic, social, and cultural needs.

With the end of World War I that freshly articulated twentieth-century national culture took its place as a full partner in the family of major nations in the world. No longer could a limited and partial national government serve the nation's needs and those of its citizens as had been the case in the early and middle nineteenth century. Now Americans and their rulers shouldered new responsibilities in trade and commerce all over the world. And they faced the challenges from new issues and problems in society, especially the movement of populations to cities from rural areas and the decline of rural areas as centers of prosperity. Increasingly in the 1920s and especially in the 1930s, there were serious problems to be solved with regard to the economy, social structure, and intergroup relations, especially between labor and capital, whites and nonwhites, the American-born and the foreign-born, but even between city and country folk.

Public officials turned to the experts in research and land grant universities, in the foundations, in the professions, and in the large corporations for encouragement, advice, and proposals to solve problems. What was the state of available knowledge about society? they asked in many different ways. It was not as though there had been no social knowledge before the 1920s or even no social sciences. As compared with the older physical and biological sciences, however, many in business, the professions, and the sciences had questioned whether the social sciences could really benefit anyone. According to these critics, the social sciences were poorly organized, lacking in rigor or scientific method, and chock-full of personal opinions rather than laws of nature and society. During the pre–World War I Progressive Era there were social scientists, many of them advocates of social reform who gathered social information to be transformed into social knowledge capable of making society a better place to live—with better public health, more prosperity, fewer social conflicts among and between social groups, more education, and the like—and whose research and policy accomplishments created the beginnings of organized social science, at least at the level of raw data and false starts that need not be repeated. By the early 1920s some social scientists, notably economists, public health officials, and physical anthropologists, were beginning to use statistics to analyze, and not merely describe, social patterns. By the middle 1930s this had advanced to other social sciences, especially sociology and psychology, and economists were on the brink of deploying advanced mathematics to construct economic theories that could predict economic events no less surely than an animal psychologist could make predictions about what gorillas or horses or cattle might do. Thus did the social sciences "go to Washington," as bits and pieces of disciplines, sometimes as ways of doing things even more than as bodies of knowledge about well-defined social phenomena. Whether by chance or not, the migration continued in the 1930s and 1940s; if we can look back at the yellowed pages of social science journals or the faded pictures of social scientists in those decades and think of what a long time ago that was, that it was literally a different world, we will begin to realize the importance of our topic. The social sciences have played an important role in national politics and policy in the twentieth century and will likely do so in the twenty-first, perhaps unrecognizably so from the perspective of, say, the mid-twentieth century.

One of the large differences between that time and our own, as our authors make clear, is a radical shift in social outlook, from a conception of a world of groups whose behavior and members are ruled by their own past experiences to that of a universe of a social order in which all individuals have autonomy, choice, and the rationality to pursue their own needs in a responsible manner; from a tightly controlled network, or system of systems, to an atomistic society of individuals who insist on individual choice in everything imaginable in art, commerce, personal relations, religion, and consumer choices, among other things. The other large difference, which is related to the first, has to do with the de-legitimation of the moral authority of science, technology, and expertise since about 1950 in American society and culture. Starting as questions in the philosophy of science, as in reactions to midcentury positivism, this antiauthority and authoritarian movement (sometimes without differentiation between the two) has fueled the so-called culture wars about science and what objectivity, rationality, and truthful merit it may or may not possess. As will be quite clear from all of the essays, these questions have indeed roiled the waters sustaining the scientific enterprise, including that of the social sciences and national politics and policy. Whether or not we have found one of the main themes of the story of the social sciences and Washington remains to be seen, but these themes have certainly been evident in other areas of American life in the last fifty or sixty years.

Part I

The Social Sciences Come to Washington

WE TURN FIRST to considerations of origins, of advent, of beginnings. It is not the case, to be sure, that the social sciences came to the nation's capital only in the era of World War II and beyond. As with many other European cultures, whether in the Old World or the New, the social sciences, or contemporary versions of them, existed from the time there was a federal government. In addition to the statutes and legal decisions of the early Republic, or even economic plans such as Alexander Hamilton's brilliant *Report on Manufactures* (1790), there was social science installed as the Census, which did not become a permanent government bureau until the early twentieth century and was throughout the nineteenth century a survey of population, economy, and society revived every decade for enumeration and explication. And there were other instances of creating social knowledge in the nation's capital and with its distinctive institutions, including ethnology, physical anthropology, educational opportunities and circumstances, military strategy and tactics, economic entomology, among many others. Clearly in the nineteenth century such federal institutions as the Smithsonian, the U.S. Department of Agriculture, the U.S. Patent Office, and the Library of Congress contributed much to the manufacture and distribution of social knowledge.

Until the international crisis of the rise of totalitarianism in the 1930s and early 1940s, the federal government hired scientists to work within governmental institutions. In 1940, however, as the international crisis deepened and the government's investment in scientific research and development increased so dramatically, a new system grew up in which federal agencies negotiated contracts with universities, laboratories, factories, and the like to produce the goods and services essential to the war effort. The new arrangements neatly sidestepped certain bugaboos of the U.S. Congress, such as too many federal employees. Waste and inefficiency continued, indeed, spiraled upward; but in the final analysis senators and representatives discovered, sometimes to their delight, that such inefficiencies of economy sometimes benefited a constituent whom they represented, and horror at cost overruns faded quietly away.

The social sciences came to Washington, then, in the 1940s and beyond, in most instances from the research and land grant universities where they had been nurtured in the later nineteenth and early twentieth centuries. War and the threat of war made the social sciences, like the natural sciences and engineering, important and meaningful to federal officials. Under the threat of invasion, defeat, and, starting in 1949, atomic warfare, the social sciences came to serve the nation's defense and augment its well-being as a society.

In Hamilton Cravens's essay on the social sciences and how they supplied the rationale for affirmative action, the transit of ideas was tortured indeed. That notion, which President Lyndon B. Johnson first enunciated in his famous address at Howard University in June 1965, referred to a deforming legacy of hatred, the product of intransigent white racism directed against the nation's African American population, that made the possibilities of personal liberty and opportunity far more difficult for the nation's black citizens than for any other group in the nation's history. Therefore, insisted the president, the nation must take affirmative steps to equalize circumstances so that blacks and whites can compete for society's blessings on an equal basis. Only in that way will the nation's wounds be healed and made whole. Only then will we have a just society. One would suppose that this message came from the famous Chicago school of sociology, the Robert Ezra Parks, the Ernest W. Burgesses, and the Louis Wirths, who wrote about conflict and assimilation in America's urban social systems. Actually, the message of cultural deprivation and oppression came not from the white Chicago sociologists, who accepted racial segregation as a fact of nature and society, but from those black sociologists, at Chicago and elsewhere, who spoke up for blacks in the 1940s and 1950s. And, in important ways, the most important contribution that President Johnson, his own intellectual retainers (especially Daniel P. Moynihan), and his academic mentors made was the importance of understanding African Americans as a group of citizens, not as an endless number of individual petitioners. In other words, communication did not always flow from Washington to the country; here the country, or at least the academy—including not simply the University of Chicago but certain other academic centers, some of them historically black institutions—served as the originators and conduits of the new ideas of psychological and social impairment rather than legal and political oppression.

Illustrative of the boundaries or limits in which the social sciences would operate once in the nation's capital is Michael Bernstein's cautionary tale of economists' efforts to bring prosperity to the nation's citizens in the several decades following World War II. The 1950s and 1960s seemed an ideal time to experiment with economic theory so as to bring about endless expansion for the nation's economy. As compared with the other major belligerents of World War II, the United States was blessed. Its economy was physically intact and organizationally and technologically robust. And it was a creditor nation to the world. Thus, even with the three recessions of the 1950s, the United States seemed to have a limitless future. Economists at the Council of Economic Advisers, building on a decade and a half of modern Keynesian theory, overrode various political obstacles to the Revenue Act of 1964, which spurred the economy, and then started the famous War on Poverty, which would go hand-in-hand with the new policies of civil rights and affirmative action to transform America, within a generation or so, into a truly free middle-class society with the remnants of institutionalized racism in fast retreat. So President Johnson dreamed. And, as Bernstein reminds us, so came the Indochinese war, the war that the president must win along with the War on Poverty. Always in such matters, national security and military considerations outweighed those of society, economy, and domestic polity. And so the nation remains, with poverty worsening and the dilemmas of the 1960s still not addressed.

Chapter 1

American Social Science and the Invention of Affirmative Action, 1920s–1970s

HAMILTON CRAVENS

A New Departure in Public Policy

On June 4, 1965, President Lyndon B. Johnson gave the commencement address at Howard University, the federally sponsored historic black college. In the preceding decade Americans had become increasingly aware of the civil rights movement and of the injection of the issues revolving around civil rights for black Americans into the national public discourse. President Johnson took a new angle in attacking discrimination against black Americans. Instead of focusing on the political and legal aspects of Jim Crow legislation, or on the constitutional struggles for civil rights in education and voting, or on the plight of black Americans in the South, he spoke—with great passion—about the social and economic circumstances of African Americans throughout the nation, including those trapped in the large urban ghettoes in the Northeast, the Middle West, and the West. "In far too many ways American Negroes have been another nation: deprived of freedom, crippled by hatred, the doors of opportunity closed to hope," he argued.

This indeed was the president's thesis. The time had arrived for extending the civil rights revolution from issues of legal discrimination and equality to those of social and economic opportunity and equality. Legal and constitutional freedom was one thing; opportunity was something else. It was time that the gates of opportunity be opened. The nation's twenty million African Americans had to be given the same chance as every other American to live with the same choices American society offered all other citizens: to become educated; to be citizens, workers, and human beings; and to pursue happiness. Johnson insisted that blacks had always constituted another nation within the American nation. Ever since the days of slavery they had been identified by whites as an inferior race or group, stigmatized, segregated, and not permitted to enjoy the full fruits of American citizenship. African Americans have constituted a *group* or *class* of persons who have been denied many elementary freedoms, many ordinary chances in life. In short, they have been victims of American society. Such victimization must be rectified, by government action if necessary. This historical burden must be cast aside. It was now the nation's obligation to right these many wrongs against this race, this group—black Americans. In arguing that, Johnson took the radical step of proposing that African Americans constituted a group or category of Americans who were entitled to first-class citizenship within the nation's political system. Now, finally, in a manner not unlike all previous stigmatized groups, American blacks were to be players in the nation's politics and governmental programs, as members of an interest group, which was the key to participation in the mainstream agendas of national, state, and local public policy.

President Johnson executed this maneuver with true intellectual and ideological artistry. He did so in large measure when he inverted the traditional American explanation for the inability of African Americans to keep up with white Americans. Instead of repeating the popular myth that blacks constituted an inferior race as compared with whites, Johnson took a large piece of contemporary social science as the bedrock assumption for his speech. All people, regardless of their racial ancestry, were born with the same range of abilities. Ability, however, was not merely a matter of what one had at birth. Ability was influenced, for better or for worse, by one's family, neighborhood, school, "and the poverty or richness of your surroundings. It is the product of a hundred unseen forces playing upon the infant, the child, and the man." In a word, biology provided potential. Society determined how things would turn out and which, if any, opportunities would present themselves.

The president then noted another—and relatively new—finding of contemporary American social science, a mystifying and disturbing paradox. In the previous generation, the black middle class had been growing steadily, closing the distance between themselves and middle-class whites. In the same period, however, most black Americans had found their lives becoming progressively worse, despite the civil rights movement's legal and political victories in recent years. Unemployment in 1930 was about the same for whites and blacks. Thirty-five years later the unemployment rate for blacks was twice as high as that for whites. Two and a half decades ago white teen-agers had a higher unemployment rate than black adolescents, whereas in 1964 23 percent of blacks were unemployed as compared with only 13 percent of whites. And so it went with other such comparative measures; thus the rising poverty Michael Harrington declaimed in *The Other America* (1962). Johnson argued that a black proletariat was forming, whose fortunes were worsening, at precisely the same time that the black middle class was catching up with the white middle class. Hence the paradox of some progress toward equality for the nation's small but growing middle class but disaster for the black masses. It was all the more dispiriting, Johnson insisted, because the larger society had become much more prosperous since World War II.[1]

In sum, President Johnson argued that most African Americans were still victims of structural inequality. For this there were several reasons. They were imprisoned in "inherited, gateless poverty." The consequences were disastrous. African Americans lacked training and skills. They could not escape the urban ghettoes in which they resided. They lacked decent medical care. So did many whites, of course, but black poverty was not white poverty. The "devastating heritage of long years of slavery" and a "century of oppression, hatred and injustice" made the difference, even though poverty for blacks and for whites had many of the same causes and cures. If these differences flowed not from race but rather from social prejudice, stigmatization, and oppression, then it followed that blacks could not completely free themselves from their oppressed position in American society. No other group in American society has had the special problem of slavery and segregation. "You do not take a person who, for years, has been hobbled by chains and liberate him, bring him up to the starting line of a race and then say, 'you are free to compete with all the others,' and still justly believe that you have been completely fair." It was not enough to open the gates of opportunity. "All our citizens," he insisted, "must have the ability to walk through those gates."

It would be difficult to exaggerate the importance of Johnson's speech, at least as a symbol of contemporary life. In that address he articulated an important shift in American political and social scientific thinking about race and equal opportunity. What came to be known as affirmative action had as its intellectual and ideological justification in the thesis that he propounded that day at Howard University: that blacks as *a group or race* have carried an unusual burden of discrimination and have been victimized by centuries of systemic, institutionalized oppression. Hence any remedies to make blacks fully participating citizens must be race based and race conscious—that is, must address their status as a group in the body politic. That burden had been placed on blacks by systemic white prejudice and socioeconomic structural realities, and it would have to be removed before they could participate as equals at the starting line of the race for attainment of society's benefits with white citizens.

What Johnson had dubbed affirmative action was therefore entirely apposite, the proper political solution to the problem of white racism as directed against African Americans. As such, it was an integral part of the national political conversation concerning race and equal opportunity. In the American political system (and, presumably, in any modern political system), public policy can only be based on group conflict and group rights, no matter what that system's political ideology and rhetoric might be. Individuals, no matter how worthy, can never be the object of public policy or participants in the creating or receiving of public policy. What has constituted public policy has been that which has affected broad categories of constituencies within the political system. This is not to say that individuals have had no influence in the political system. They have, and not merely as individual voters and officeholders. Such issues as taxation and regulatory exemptions, as well as more traditional forms of individual patronage, such as immigration and land use, have often responded to individuals rather than to groups. It is crucial to grasp this distinction in the nation's political system, for it lays bare the problems of the politics of minority groups and affirmative action in American history.

Put simply, public policy and patronage have run as totally different political processes in American politics. And this has been true for constitutional issues and resolutions as well, and the courts have responded both to suits involving categories of persons (hence public policy) and individuals (hence individual justice, or, more simply, individual patronage). It is also a truism that neither the political nor the constitutional system has bestowed the benisons of individualized patronage upon members of stigmatized minorities, especially persons of color. The color line seems to have disqualified individuals or groups from redress or relief of any kind from either the political or court system. President Johnson knew exactly what he was doing when he called for affirmative action, then; he was attempting to create a role for American blacks as a group within the larger political system so that they could compete for public goods. That was what he meant by his homely metaphor of having everyone begin at the same starting line. Of course, in American political rhetoric such group references have commonly been made as if free individuals were the only persons involved. As several works on "whiteness" have made clear in recent years, many of the European immigrant groups found that they had to "become white" before they could compete within the body politic.[2]

And where had that most pragmatic, restless, and enigmatic of contemporary politicians, Lyndon Baines Johnson, gotten his ideas? In a sense, Johnson needed no tutors on the subject of race, for he had confronted the issues of race and class in his career in Texas and national politics.[3] In discussions with his advisers in and out of government, Johnson had decided that the constitutional and legal parts of the civil rights revolution had seemingly been resolved with the enactment of the Civil Rights Act of 1964 and the Voting Rights Act of 1965. Johnson concluded that it was time to start a new phase of the civil rights revolution, to push for social and economic equality. Two aides, his special assistant, Richard Goodwin, and Assistant Secretary of Labor Daniel Patrick Moynihan, had written the speech. Moynihan himself had already made something of a name for himself in governmental and social scientific circles with the distribution of a pamphlet he had written, *The Negro Family: The Case for National Action* (1965)—the thesis of which became, in effect, President Johnson's speech at Howard University. What I am after in this chapter, however, is not how Moynihan "influenced" the president of the United States. What we want and need to know here is where did these radical ideas of Moynihan's come from, and, in particular, what was their relationship to the discourses of the American social and behavioral sciences in the generation or so preceding the Howard University commencement address.[4]

But there is more. Many scholars and commentators agree that there was a fundamental change in the civil rights movement's goals and ideology in the 1960s, and that of subsequent government policy, concerning racial equality. One can see that in Johnson's speech. Clearly Johnson wanted African Americans to be active players, not passive objects, in the political system. And just as clearly for this to occur was needed their legitimation as a group in the body politic. Yet no president—no political leader, really—could merely wave a magic wand to effect such a change. The political system would have to change so that American blacks could participate as a group that could fight for its own interests, whether in the public policy or the patronage sphere of politics and public goods. This was the only way to place blacks as a political interest group on the same footing as whites.

Since the 1960s there has grown up a school of interpretation whose champions criticize this transformation of American politics and constitutional development. Affirmative action, they insist, has distorted and twisted ancient and honorable American notions of equality and fair play. Among those who make these arguments is Paul D. Moreno, in a generally well received book on constitutional issues of equality in the workplace since the 1960s, and Herman J. Belz, in a more general work, *Equality Transformed* (1991). Before the 1960s, they argue, the civil rights movement's leaders and followers were committed to a tradition of an individualistic, meritocratic, and color-blind notion of racial equality. Since the 1960s, they continue, the civil rights and black power movements insisted that, because blacks were victims of history, they deserved preferential treatment over whites whose qualifications made them more deserving of the public goods for which they were in competition, such as college or professional school admissions, jobs in the public sector, and the like.[5]

The question arises whether this is a valid historical interpretation of the recent past. Has that tradition of a meritocratic, color-blind, and individualistic notion of equality of opportunity actually existed as historical fact? It is certainly true that in the 1950s the rhetoric of the national debate over equality and civil rights was cast in such a discourse, with such meritocratic "may the best qualified individual win" and other so-called color-blind themes. What that actually meant, however, may well have been a vastly different thing. And if blacks began to consider themselves a group in the body politic, it is hardly surprising that at least some may have insisted, most likely for tactical reasons, that they deserved special treatment, given the special resonance then of the national discourse of victimization. Indeed, a careful examination of American politics in the last several decades would reveal that such rhetorical claims were hardly restricted to African Americans.

Here an examination of the social and behavioral sciences' discourse on race and equality from the 1920s to the 1970s can offer us a new perspective, distinct from that of the legal and constitutional historians. After all, the law is pretty much limited to a case-by-case analysis, which in turn yields, or can yield, an individualistic point of view, whereas the social sciences can offer that or perspectives that view problems from a more remote and impersonal perspective, emphasizing groups and institutions. It has been a convenient social fiction in the American political tradition to assume that American constitutional law and its discourses are oriented toward individuals, not groups, but in fact individuals in said constitutional and legal discourses have always "stood for" the groups they have "represented" in the larger society for matters of public policy (as defined above). That individuals seek redress through the judicial system does not obviate the fact that our political system deals with groups, never with individuals, in the formation of public policy; to the extent that such issues as affirmative action become political as well as narrowly legal issues, they must involve (and be oriented toward) interest groups in the body politic. Belz and Moreno ignore that whether or not there was an individualistic, meritocratic work ethos in place, it could only be an ideology for a group or a category of people, not for an individual, save in a particular lawsuit (which can only have consequences for an individual); indeed, in any event, the so-called Protestant or middle-class work ethic has always been a credo of individual conformity to group standards or norms.[6] With that in mind, let us turn to the history of the social science conversations and discussions on race in American life in the half century or so before affirmative action became governmental policy. This may enable us to think clearly and freshly about the problem of group and individual in American politics and political culture.

The Chicago School and the Concept of the Group

Clearly the concept of the group has been an integral element in American social thought and American social science discourse since the late 1830s and early 1840s, when the group was first recognized in American social thought and in the social disciplines as the basis of American democratic society. Indeed, the invention of the group as a category of analysis owed much to the democratic revolution then taking place in antebellum American society and culture. Hence it is monumentally counterfactual to say, as Thomas Haskell has in his widely cited monograph, *The Emergence of Professional Social Science* (1977), that group analysis arose only with the recognition, in the late nineteenth century, that American society was interdependent. That sense of interdependency, along with the discourse of group analysis, was clearly evident in the later 1830s and early 1840s to any scholar who wished to peruse either the relevant original sources or the rather voluminous secondary literature on the democratic revolution of the 1830s and 1840s in American culture and society.[7] Nor is it necessary to examine the sources and accounts of intellectual history to recognize this point. There were also changes in the character of American institutions, such as the churches, the political parties, the public schools, the corporation, the identification (and often stigmatization) of distinct groups in the national population from the 1830s on, which indicated that the group, not the individual, accounted for how things now worked.[8] In other words, group and individual are independent units of description and analysis in both social thought and action in the larger culture as well as in the social disciplines themselves.[9] To say that there is a tradition of individualism is to misapprehend the subtext. Usually such traditions have been pleas for group conformity—the so-called Puritan work ethic—as in classical laissez-faire economics or in the novels of Horatio Alger, not actual analysis of the individual and society. Much the same is true of our so-called frontier tradition of individualism; that which was genuinely individualistic or apart from society, and thus from groups, such as the fur traders or the mountain men of the early-nineteenth-century frontier, can be situated in the period before the late 1830s. And, indeed, perhaps such individualism of the frontier was only so much myth.[10]

From the 1920s to the 1950s, then, the social sciences had a distinctive model of the taxonomy of social and natural reality—of the order of things, as Michel Foucault would have it—that was both an exact diagram of the relations between the whole and the parts and of the group and the individual. According to this model, the whole is greater than or different from the sum of the parts. Each and every part is interrelated with every other part. Thus University of Michigan sociologist R. D. McKenzie and his associates, in *The Metropolitan Community* (1933), outlined the basic changes in the nation's cities since the advent of motor transportation—and exemplified that model of reality. McKenzie and his colleagues thought in ecological terms, as was the fashion of that era. The nation's cities constituted a network of communities, differentiated from one another in a variety of ways. There were the super communities, the relatively small number of colossal urban centers; then there were other cities that had particular economic or trade functions. But it was no longer a matter of cities attracting population from the countryside and abroad to work in factories. Thanks to the changes that had occurred since the First World War in the functional reorganization of the interrelations of institutions and services within the city and its environs, a new kind of city was emerging. "The old communal pattern, characterized by compactly integrated cities which were sharply differentiated economically and culturally from the surrounding settlement," McKenzie argued, "is being replaced by a more open regional community composed of numerous territorially differentiated, yet interdependent, units of settlement."[11] The whole, then, was an aggregate, an ecology, that is, in which the parts were interdependent, and a change in any one could have incalculable consequences for all other parts of the whole. The whole was dynamic, not static; fluctuating, not finished. Its causes were to be found in the present as well as in the past, and, in a word, oil and water did mix: qualitatively different elements went together in the larger whole. This was a very conservative perspective, to say the very least.

In these interwar years, then, the University of Chicago, McKenzie's graduate school, became famous in sociology, especially in urban sociology, social psychology, and in the application of statistics to social data. As such, it became the major center of social science thinking and research on issues of racial and ethnic group competition, the structure and survival of the family, and other matters central to what would become affirmative action. Among the leading lights there was Robert E. Park, a former newspaper reporter and aide of Booker T. Washington. Primarily interested in developing a school of urban sociology, Park believed sociology should be a positivist social and behavioral science with its own natural laws and generalizations about human nature and society regardless of time and place. Park had studied in Germany, where certain philosophers of science taught him to distinguish between history and science. Historians studied events—specific, particularized, individual, and unique—whereas scientists (including social and behavioral scientists) studied the typical, recurrent, and general objects, which they could define conceptually and assign to classes, types, and species, so that they could draw general conclusions, which historians perforce could not. Park insisted that personalities, groups, institutions, societies—each with their own subcategories, characteristics, and typical modes of change—were as evident in human nature and society as they were in nature. Park had no objection to the use of historical research provided it served the sociologist's purposes of uncovering those laws. Park saw in the city all the problems and social forces driving the evolution of society and civilization: it was the dominant arena for social change; it was the progenitor of society itself; it was a convenient laboratory for sociological research; and the rapidity of social change in cities caused social problems, the study of which would be helpful to reformers.

Park and his associates believed that the city's spatial organization reflected the capitalist market, with its exchange of goods and services. The city's ecological configuration showed how competition sorted out where groups of people lived and worked. Competition had other effects as well. Since different groups had widely varying social power, inevitably there were winners and losers in land values, social rewards, public goods, and the like among human groups, whether those groups were classes, races, ethnic groups, religious groups, or whatever. Competition produced what Park and his associates thought of as "natural areas," which were differentiated and specialized for the city's various groups for living and working. These natural areas resulted not from design or from a plan but rather from the tendencies inherent in human nature and urban structures. "What has been called the 'natural areas of the city,'" Park wrote, "are simply those regions whose locations, character, and functions have been determined by the same forces which have determined the character and functions of the city as a whole."[12] Hence scientists—in this instance urban sociologists—could study a city's natural areas and its larger ecology just as chemists, physicists, and zoologists could study their objects.[13]

Park and his colleagues studied ethnic problems in the city extensively. William I. Thomas actually started the Chicago school's interest in ethnic and racial studies with his pathbreaking study, done with Florian Znanecki, *The Polish Peasant in Europe and America* (1918). They insisted that the first generation of immigrants found life in their adopted country destabilizing, but, in the long run, their children were assimilated into the larger culture, which showed that the entire cycle of race or group relations was positive—it had a happy ending. In that sense, Thomas wrote the archetypical Chicago study of racial conflict and assimilation: competition led to good things. It was a script that the city's chamber of commerce would have loved—and could have written, probably with more pizzazz. Thomas had come to believe that all races were intellectually equal—a most unusual position for a white person then, to say the least—and so he saw ethnic conflict as having its final resolution in the differences among individuals, not groups, providing for balance and equity. But that resolution would come in the very distant future, he believed, long after the final stage of total assimilation of all groups in the society.[14] Thomas had to leave the university in 1918 after he was accused of improper sexual conduct (the charges were later dropped), and the intellectual leadership he had exhibited at Chicago now passed to Park and others.[15]

In 1921 Park and Ernest Burgess published *An Introduction to the Science of Sociology*, in which they outlined their school's main ideas. None of the Chicago sociologists, like virtually all social and behavioral scientists in the interwar years, could conceive of the individual as standing apart from the group to which "nature" or "society" assigned him or her.[16] Human nature could only be acquired through interaction with other persons; it was a group nature. Individuals could not exist apart from one another or the larger society. People in a society and community interacted with one another, and such interactions were either primary or secondary. Primary interactions meant those of whole personalities and face-to-face contact. Such were typical of primitive or rural societies. Secondary interactions took place in a modern society. They were rationalized and partial. Inevitably there was instability in such societies because of group competition and conflict.

From Continental sociology and social thought, the Chicago sociologists took the idea that the theme of modern history was the movement from community to society, from gemeinschaft to gesellschaft, from rural or traditional peasant societies with primary social relations to modern societies with secondary relations and conflicts. In time conflict turned to accommodation, as competing groups worked out their interactions. Most of the Chicago sociologists remained quite complacent about such conflict and accommodation. As true white liberal optimists, they assumed that the ultimate end of group conflict would be assimilation. Their African American associates, however, notably E. Franklin Frazier and Charles S. Johnson, were far less optimistic. They knew full well the harsh realities of white-black interactions in North America—from slavery to segregation and white domination of blacks. Assimilation was simply not a realistic possibility to their way of thinking.[17]

Louis Wirth's pathbreaking *The Ghetto* (1928) is a good example of Chicago assimilationist sociology. What began as his investigation of Chicago's Jewish ghetto soon led him to examine the natural history of an institution and the psychology of a people, going back to the early history of the Jews in Roman times and the Diaspora, or dispersal from their homeland; to the life of European Jewry in the Middle Ages and the anti-Semitism of the Christian Crusades, which created the first ghettoes in European cities in the fifteenth century; to a discussion of the ghetto of Frankfurt, Germany, which Wirth analyzed as a typical ghetto, very cosmopolitan, with much Jewish solidarity. Were the Jews a race or a type? They were both, created as such by the forces of the past and the present, adapting to the pressures and opportunities around them. Wirth also noted the differentiation of European Jews into a western European urban type and an eastern European village and rural type—the former cosmopolitan, educated, mercantile, and adapted to modern civilization; the latter a farming and rural population, highly insular, very traditional—so that there was not one Jewish type of mind but several.

Wirth devoted the second half of his book to the history of Jews in America, from colonial times to the contemporary Chicago ghetto. That ghetto's founding was "fairly typical of what happened in the last one hundred years in every urban center in the United States," he noted. Initially the Jewish community was scarcely distinguishable from the rest of Chicago. As its population increased, however, the European ghetto's typical community institutions gradually crystallized, and the "addition of diverse elements to the population" resulted in "diversification, and differentiation, and finally in disintegration," Wirth concluded, thus applying the cyclical model of group or ethnic relations of the Chicago school to the history of Chicago's Jewish ghetto. But as the Chicago ghetto swelled its numbers, more and more members adapted to the majority Gentile culture. They found it was easier to live outside than inside the ghetto. After the Great War the ghetto went into decline. But assimilation could only go so far, and in the last phase of the cycle—a cycle as natural and inevitable as that of any animal or plant species—when Chicago's Jews found themselves again excluded from intimate association with Gentiles, they recoiled and rediscovered their identity as Jews in their own middle-class suburbs, synagogues, country clubs, and the like. The cycle continued because European anti-Semitism became manifested in Chicago, as, for example, in the antagonism between the Polish and the Jewish immigrants within the ghetto and the Anglo-Saxon middle class beyond the ghetto. The modern ghetto was based on New World sentiments and group antagonisms rather than on Old World fences and borders. "Interaction is life, and life is a growth which defies attempts at direction and control by methods, however rational they might be, that do not take into account this dynamic process," Wirth said. "In the struggle to obtain status, personality comes into being. The Jew, like every human being, owes his unique character to this struggle."[18] Ultimately Wirth and his colleagues thought that full assimilation would occur according to the natural laws of society, not the wishes of individuals living in that society.

Like sociologists and social scientists elsewhere in America, Chicago sociologists grounded their ideas on the existence of competing groups in society. Individuals were members of groups. All their ideas and actions resulted from that root assumption—and reality. Chicago sociologists were also implicitly evolutionary and optimistic. Park and his associates blithely assumed that cultural development was progressive. So long as they were dealing with white European groups, their ideas had some plausibility. Their entire system of ideas broke down, however, when they confronted the problem of African Americans in American culture. It is instructive to compare how Park and two of his students, Edward Byron Reuter and E. Franklin Frazier, faced the issue.

On the whole, the Chicago sociologists never developed clear conceptions and definitions of what constituted the majority culture or what assimilation actually meant. Park's views illustrate the point. At first, in 1913, Park declared that assimilation was a process whereby groups with distinctive characteristics entered as coordinate parts into society's practical working relationships while retaining their distinctive group peculiarities. He and his colleagues compared the circumstances of European peasants with those of African Americans. As a devoted champion of racial equality, Park assumed that, as European peasants had passed from "community" to "society," so, too, would American blacks.

By the late 1920s, however, Park had serious misgivings about this "melting pot" theory of social evolutionary progress. He now believed that slavery had stripped blacks of any important or meaningful survivals of their African culture. European immigrants, on the other hand, sustained a rich tapestry of their Old World culture. Using the Chicago assimilationist model, one could argue that European immigrants had moved from a culture, a rural society with a moral order sustained by mores and folkways and values implicit in ritual and tradition, to a civilization, an urban society in which secondary contacts, such as the capitalistic marketplace, work, rationality, and the modern state, defined interactions among members of different ethnic groups and tied them together. Whatever culture American blacks had was Anglo-American culture but in an obviously crippled form. In a word, segregation and slavery made assimilation impossible for African Americans; the comparison failed, and with that, so did Park's fond hopes of ineluctable assimilation and progress.

Reuter's experiences illustrated the same problem with the assimilationist model from a slightly different perspective. He grew up in rural Missouri with rather more than the typical rural white prejudices toward blacks. After all, Missouri had been a slave state, and after Appomattox segregation became the law of the land. This southern cultural background influenced Reuter considerably. At the state university in Columbia, he learned his ethnic sociology from Charles Ellwood, who defined social progress in tooth-and-claw Darwinian metaphors. Reuter found nothing to contradict white racism here. Slavery had permitted African Americans to retain their ancestral tropical laziness, as well as the whites' language and the rudiments of white culture. But American blacks had not yet learned the self-mastery essential to being responsible—meaning white—American citizens. In time, Ellwood argued, natural selection would alleviate the problem, as blacks learned to work efficiently and to own property as whites did. Ellwood was unclear whether race or culture caused black inferiority. Here was Reuter's opportunity. When he arrived at Chicago in 1914, he dedicated himself to the study of mulattoes, persons of both black and white ancestry. Thus he could examine the effects of race mixing and, by implication, whether natural selection would lead to racial improvement and social progress. Reuter easily adopted the notions of Chicago sociology; apparently virulent racial prejudice was no barrier to embracing Chicago sociology, which, given the pervasiveness of white racism in American society then, was hardly surprising.

In *The Mulatto in the United States* (1918), Reuter discussed how and why mulattoes defined race relations. He insisted, often with scant evidence—gross tabulations from the census as "objective" racial data, really—that mulattoes, the products of miscegenation, dominated the leadership class of American blacks. For Reuter, comparisons of similar situations worldwide confirmed that fact. So far Reuter had simply upheld a common conceit of white supremacists, that high-achieving blacks always had some white racial ancestry. Yet Reuter had a problem—a professional one. If he were to persist in his frankly racist views, his prospects for a good academic career outside the South were doomed, for the world of professional social science of the post–World War I era demanded that all don the mantle of scientific objectivity and abandon extreme views, whether of the Left or the Right.[19] Peer pressure for ideological conformity and academic professionalization seemed to work hand in glove. Reuter learned quickly. He eliminated all expressions of overt racial prejudice. After several temporary appointments, he went to the University of Iowa and remained there for most of his career, becoming noted as an important scholar of race relations. His solution to the racial problem was for blacks to develop a full-blown racial nationalism that would give them the stability to become established in American society.[20] This was segregation, not assimilation; Reuter thus rejected the Chicago assimilationist model.

E. Franklin Frazier saw racial matters quite differently. He came from an upper-middle-class black family in Baltimore. Thus he identified with the institutional and educational conservatism of that social class. Yet as a member of a black elite competing with (and losing to) a new black bourgeoisie innocent of that elite's cultural values, Frazier found himself increasingly alienated and isolated within American national culture. At the same time, he strongly believed in an American version of socialism, presumably as the solution to racism and segregation, and he emphasized how the dominant economic classes and interests ruled politics, society, and economy. His white colleagues at Chicago, teachers and students alike, accepted neither perspective. He found himself ideologically isolated in graduate school. Initially he accepted the Chicago school's basic ideas, as any conscientious and ambitious graduate student would. In particular did he embrace Park's comparison of black movement from plantation to urban ghetto with the experiences of European immigrants to American cities. This enabled him to apply all the Chicago school's theories to black-white issues.

By insisting that the family was the fundamental building block of society and its class system, however, Frazier eventually took a different intellectual path than his colleagues. In his doctoral dissertation, published as *The Negro Family in Chicago* (1932), as well as in *The Free Negro Family: A Study of Family Origins before the Civil War* (1932), Frazier insisted that slavery had stripped American blacks of their African cultural heritage—including, above all, their familial structure. What remained were the rudiments of the Anglo-American culture. Crucial for African Americans coming out of slavery, then, was the extent to which they had been able to absorb the whites' culture. Those with the best opportunity to do so were the free blacks, who could most readily develop proper habits of hard work, sobriety, family cohesion—including a two-parent household—and a variety of other such characteristics typical of the white middle classes. "These families have been the chief bearers of the first economic and cultural gains of the race, and have constituted a leavening element in the Negro population wherever they have been found," concluded Frazier. "We have noted how frequently the descendants of these families are still found today in conspicuous places in the Negro world."[21] Put another way, the slave family yielded what he dubbed the natural family; here authority was denied the slave parents and spouses and reserved for the white masters, with disastrous consequences. This natural family was but an adaptation to the slave system. Emancipation revealed its glaring flaws: a lack of parental authority and its consequences. The alternative was what Frazier called the institutional family, an adaptation of the middle-class white family, with a legal marriage, property ownership, two parents in the household—the so-called nuclear family. Because initially he wanted the Chicago assimilationist model to be valid, Frazier ignored the nuclear family's fate since emancipation, which was disintegration and divorce, because it possessed no other institutional or cultural support save the naked marriage contract itself. Yet in time he would change his tune.

In his major work, *The Negro in the United States* (1949), Frazier mapped out the history of white-black relations in America organized around the race relations cycle. It was not Park's version of that cycle that he deployed, however, but that of Emory Bogardus, another Chicago sociologist, who was always pessimistic and who insisted, finally, that the outcomes of Park's race relations cycle were the mutually exclusive ones of assimilation or segregation, thus making a distinction between ethnic and racial competition. Frazier took up this theme, using the black family as his point of departure. He agreed that since World War I blacks had achieved some of the status of a white immigrant ethnic group, but he saw that whatever social and economic progress African Americans had achieved had come about not as the result of increasing white tolerance toward blacks or the diminution of white attitudes as institutionalized in the larger society but merely as the consequence of large and impersonal socioeconomic changes in the structure of society, such as the demand for black labor during wartime leading to greater employment. Unlike Park, who naively assumed that whites would relent in their sense of superiority over blacks, Frazier assumed that social conditions and the political necessities the federal government was faced with in World War II would alter power relations among the races, with some benefits for African Americans. Frazier's was thus a bleak view of the race relations cycle, in direct contrast to Park's. If it is true that in his later years he lost some of his optimism about race relations, Park apparently never considered that blacks, like other nonwhite groups, were denied their fundamental rights as citizens, something that did not happen to European immigrants, on the whole, and which in any event his assimilationist model failed to take into account.[22]

Race: Culture and Personality

The Chicago school's distinctive contributions to the evolving discourse on race in American culture cannot easily be dismissed. Its members reintroduced, for their era, notions of group conflict and competition. They gave them a historical dimension, thus implying that there was a causal relationship between past and present. They argued furthermore that although ethnic and racial groups changed over time, they did possess particular characteristics at any given time—or during any particular historical epoch, for that matter. In the long run, it was the school's most brilliant scholar and dissenter, Frazier, whose work on the black family had the largest impact of any Chicago scholar on the fledgling behavioral and social science field of black studies. In 1935 the American Council on Education, a Washington, D.C., lobbying group in the field of education, which also acted like a trade association, established the American Youth Commission. In turn, the commission oversaw several dozen research projects on the needs and problems of American youth. In 1938 its members sponsored four studies of how black youth in contemporary America faced distinctive problems—that is, racial problems—in their development as individual personalities. The social and behavioral scientists thus recruited constituted some of the nation's leading black scholars, such as sociologist Charles S. Johnson and psychologist Allison Davis, and also major white researchers, including psychiatrist John Dollard and sociologist W. Lloyd Warner.[23]

Frazier was part of this group. His *Negro Youth at the Crossroads: Their Personality Development in the Middle States* (1940) was a comparative study of black youth living in Washington, D.C., and Louisville, Kentucky, two "border state" communities. He investigated the impact of the crippled black family and of segregation on the personalities of black youngsters. He also outlined a new theoretical departure in the social scientific discourse on race in America. Frazier forged causal links between slavery and segregation. Slavery's legacy made the history of African Americans unique—and very distinct from that of white European immigrants. The destruction of the family in slavery, a result of the master's unchecked power, led to the slaves' systemic victimization in post-Appomattox times. Segregation and victimization imposed a multitude of social pathologies upon African Americans, which in turn fed the fires of white racism.[24]

Frazier argued that these young people were in transition from a southern and a rural way of life to a border state and city way of living—from "community" to "society," in mainstream social sciencese. But problems abounded. They were rooted in family patterns from slavery to emancipation and "self sufficiency." A light skin and money historically had been the keys to high status within the African American community, Frazier insisted, but in this century one's occupation has mattered, too. Half of Louisville's blacks were common laborers; only a third of Washington's were. Common laborers had to live in unspeakable conditions in slums and struggled to stay employed at menial wages, which in turn undermined parental authority within the family and caused a vicious cycle of social problems. Middle-class blacks had, on the whole, more stable lives. Their families were healthier than those of the black masses. They held skilled or semiskilled jobs if they were men and worked as domestic servants if they were women. They belonged to churches and lodges and often lived in neighborhoods that, although rigidly segregated, were relatively free of disorder and crime. Between the middle class and the upper class were the less successful professionals, owners of small businesses, and post office employees. The upper class was a small group of professionals, educators, ministers, physicians, and lawyers, together with the most successful businessmen and their families. The black community as a whole was segregated from the larger white society, which devalued the black social class system. Each strata among blacks was a rung lower than it was among whites—the black upper class was equivalent to the white middle class, and so on.

Frazier examined how the family, the neighborhood, the school, the church, the job market, and social and political movements influenced the youth in his sample. What he found disturbed him. The family, as the fundamental institution, determined class. Family life shaped personality and the individual's life course. Institutionalized white racism outside the community affected and damaged personality development within each and every neighborhood and family. Black children grew up very aware of white hostility to them as a group, regardless of their own individuality as persons. Within and without the black community, what was good and highly prized was being light skinned; darker hues were of lesser value. Even in the public schools, which had far more dedicated teachers and better funding than black schools in the Deep South, the blacker the student, the less attention he or she received, thus showing the influence of white racism even on the thinking of African Americans and their treatment of one another. And so it went in the church, the job market, and social and political movements: what was white was better and to be preferred over what was black. Frazier concluded his book with extensive interviews with a representative boy and girl, the one middle, the other working class. The middle-class boy believed that blacks could compete with whites if given a chance, whereas the lower-class girl was too demoralized to have any hope whatsoever. Institutionalized white racism had various effects; all African Americans were victims of white racism and discrimination in varying degrees.[25]

Frazier's colleagues in this group agreed: institutionalized white racism distorted personality development among African Americans. In their study of black personality and color of skin among Chicago blacks, University of Chicago social scientist W. Lloyd Warner and his associates argued that "color becomes more acute and painful in its consequences" for those African Americans who "most completely accept traditional American values." The more able, sophisticated, cultured, and refined a black person became, and the more that person accepted and represented the nation's political and social ideals, "the more serious is he made to feel that his race, and race alone, bars him from enjoying the full rights of American citizenship." This cruel paradox was perhaps the greatest challenge to "the national faith in the supremacy of individual merit." The significance of color, then, had as serious implications for the dominant race as it did for the minority race in American society, for it put "an ominous question mark after the most cherished national ideals," namely, that at least in America, anyone could rise from modest beginnings to "complete success."[26] In *Children of Bondage: The Personality Development of Negro Youth in the Urban South* (1940), the black Howard University child developmentalist Allison Davis and the white Yale University psychologist and psychiatrist John Dollard probed the difficulties and problems of black youth in southern cities, with methods and conclusions not so startlingly different from those of their colleagues. The black family, and therefore black youth, was victimized by systematic, institutionalized white racism in ways not so dissimilar from what Frazier had discovered in the border cities of Louisville and Washington, D.C., save that southern white racism was more virulent than white racism in the border areas of the nation. And in *Growing Up in the Black Belt: Negro Youth in the Rural South* (1941), Fisk University sociologist Charles S. Johnson concluded that the social structure of the rural South remained rigid, thus crippling entire generations of rural blacks who did not leave for the North or even for cities in the South.[27]

Should anyone be too obtuse to grasp the central message of these studies, a primer of a sort appeared before the studies themselves. Ira De A. Reid, professor of sociology at Atlanta University, published *In a Minor Key: Negro Youth in Story and Fact* (1940) at the American Youth Commission's invitation. He dealt with mortality, home environment, literacy and education, agriculture, the job market, the situation for black professionals, and the churches, among other topics, consistently pointing out black-white disparities. Thus the reader learns that death rates per 100,000 for black and white boys and girls from tuberculosis had seriously declined from 1911–1915 to 1931–1935 but had a racial twist, with almost four times as many blacks as whites dying in the first period and almost eight times as many in the latter period. And in seven of ten southern states almost five times as much money was spent on white as on black pupils.[28] Institutionalized white racism was causing serious social pathologies within the black community; that seemed to be the thesis that Frazier and like-minded colleagues pushed.

A later generation of Chicago social scientists approached the problems of white oppression and victimization of blacks from a somewhat different angle than Frazier. Here W. Lloyd Warner was a major figure. Warner's previous studies of social class and caste had taken the promising perspective that such taxonomies had subjective as well as objective aspects—that one's status was not merely the consequence of one's material possessions but also what persons within and without the status group thought about the behavior of the individuals within that status group, whether it were a class or a racial caste. Another important figure in this school was W. Allison Davis, Warner's younger colleague. Davis brought a different perspective to the problem—that of an African American who also understood the world of the poor as well as that of the middle class and of the world of blacks no less than that of whites. Together with colleagues in the field of social anthropology, and under Warner's direction, Davis published *Deep South: A Social Anthropological Study of Caste and Class* (1941), a study of race and class relations in a lower South community and its rural hinterland.

Based on two years of interviews within the community by Davis and his wife, and by Burleigh and Mary Gardner, black and white social anthropologists, respectively, *Deep South* provided a detailed guide to class relations among whites and blacks and then the caste relations between whites and blacks. If the general picture of a rigidly segregated caste society was familiar to those with at least a nodding acquaintance with the South, the details of class and caste relations were fascinating, especially on such matters as familial codes of conduct by class, all livened with direct quotes from various interviews.[29] In this work, one could again see the general argument that Frazier and others had advanced on white-black relations. Whites oppressed and victimized blacks. The consequence was the distortion of black institutions, beginning with the family, the central institution of black—and white—society. In other studies in the 1940s, Davis, who had now become professor of education at the University of Chicago, conducted investigations with colleagues and students in which the shaping of the personality of blacks was taken up and in other instances in which the impact of social class and caste on intelligence as measured by IQ tests was discussed. Davis attacked IQ and IQ testing from a general philosophical perspective. He insisted, for example, that intelligence was a dynamic system of functions, not a single, static, unchanging entity. He also argued that the tests measured a very narrow range of school tasks of great importance to middle-class whites, "chiefly verbal comprehension and fluency." But such things were not so important to working-class or minority persons. The tests were fallible and culture bound, then.[30]

Gunnar Myrdal: The Great White Hope

In the mid–1940s appeared the liberal Swedish economist Gunnar Myrdal's massive study, *An American Dilemma: The Negro Problem and American Democracy* (1944).[31] The Carnegie Corporation initiated and funded the project. Troubled by what they dubbed "the Negro problem" for some time, the

Carnegie trustees finally decided in the mid–1930s to underwrite a comprehensive study of the entire problem and to have a foreign scholar direct the project. They selected Gunnar Myrdal, a social economist at the University of Stockholm who was familiar with the United States and its academic and philanthropic institutions—and culture. Myrdal was also a liberal who believed that social science could improve social life. In 1938 Myrdal arrived. He coordinated the work of a small staff in New York City and that of several score of major scholars in the field of African American history and life, regardless of whether they were white or black, over the next several years. That team produced several spin-off studies that augmented the Myrdal study's influence in its field.[32]

Myrdal wrote the resulting tome, *An American Dilemma*, totaling some 1,483 pages, with two chief assistants, Richard Sterner, another Swedish social scientist, and Arnold Rose, an American. *An American Dilemma* clearly was intellectually Myrdal's product even though he had gathered information and perspectives from many American citizens and social scientists. He was shocked and appalled by the sheer extent of institutionalized racism he found throughout the United States. Segregation, he realized, was deeply implanted throughout the structures and processes of American society and culture. Myrdal was both a resolute antifascist and an admirer of America, especially of its potential for democratic innovations as compared with Europe. He insisted that the American dilemma was the nation's conflict between its creed of liberty and democracy and its treatment of African Americans. That American creed meant a profound commitment to the equality of all Americans and their rights to liberty as *individuals*. Myrdal assumed that the eighteenth-century Enlightenment was the American creed's touchstone. He argued that the Enlightenment stood for racial egalitarianism, which in turn shaped the Declaration of Independence and the American Revolution. The nineteenth century's institutionalized white racism against blacks was, he insisted, an aberration, the result of slavery's expansion. Myrdal recognized the "scientific racism" of the nineteenth century and the segregationist social practices that went with it, but he assumed that modern twentieth-century scientists—especially geneticists, biologists, and anthropologists, among others—had already invalidated these racist views with the new scientific egalitarianism, a fitting reiteration of the Enlightenment.[33] Like many economists, Myrdal used the concept of cumulative causation, insisting, in other words, that a whole host of factors were now changing the fortunes of black Americans and that while taken individually each might not have much effect, their accumulation created a dynamic situation for change in race relations.

Ultimately, Myrdal was an optimist. He really believed that the solution was at hand. It was a simple matter of educating the nation's elites who, once they had been converted to the cause of racial justice, would make the changes in society, economy, and polity to bring about progress. Ultimately, he argued, "the Negro problem" was a white problem, a problem in the white mind; it was the white community that could solve it. The war emergency, as well as the Great Depression, had highlighted that problem and brought it to the forefront. "This War is crucial for the future of the Negro, and the Negro problem is crucial in the War," he wrote. With the decay of the caste theory and the growing toleration of white Northerners toward blacks, as distinct from the white South, blacks will demand more insistently their freedom. "The Negroes are a minority . . . but they have the advantage that they can fight wholeheartedly. The whites have all the power, but they are split in their moral personality. Their better selves are with the [black] insurgents. The Negroes do not need any other allies," he continued. Thus World War II raised to a high point the problem of the color line; Myrdal suggested that the war was truly an ideological one about race relations. American race relations have become in the last few years not merely a southern problem but a national and even an international problem. "America can demonstrate that justice, equality and cooperation are possible between white and colored people," he wrote, thus laying the groundwork for the notion that America would have international leadership thrust upon it at the war's end, and the resolution of the nation's race problem would be the nation's opportunity for world leadership, for making this a better world. The social sciences in America were equipped to meet the postwar world's demands. "To find the practical formulas for this never-ending reconstruction of society is the supreme task of social science," Myrdal wrote. "The world catastrophe places tremendous difficulties in our way and may shake our confidence to the depths. Yet we have today in social science a greater trust in the improvability of man and society than we have ever had since the Enlightenment," he concluded.[34]

The influence of Myrdal's book upon the American social and behavioral sciences is difficult to overestimate. Within the general discourse of black studies, there were two arguments—one that whites, for the most part, advocated, and another whose sources were largely from black scholars. The first theory—and theorists—argued that the problem was essentially something that would be resolved when black people had the same political and civil liberties and rights that all whites did. That is, while other factors were important in causing "the Negro problem," what would resolve it was full political and constitutional leadership. The other, which really began with Frazier after he left Chicago, focused on the damage that slavery and segregation had done to the black family. It would be a silly reductionism to say that because someone was white or black that inevitably determined one's allegiance to one argument or another, for blacks and whites championed both perspectives.

But the perspectives or arguments themselves came from different vantage points, regardless of the identity of who attacked or defended them. The first was essentially a white perspective, looking at the problem from within the white community. The second was essentially an African American point of view, for its advocates emphasized that, far from political and constitutional equality being a sufficient remedy, more was needed, for blacks had been victimized by white racism since the early seventeenth century. America had always been a white man's country. All of the Enlightenment's fine promises, as well as those of the Revolution and of American democracy, applied only to whites. Both views assumed that social action occurred within and among groups: individuals were not causal factors. But the political-constitutional argument assumed that all blacks needed was the right to vote and the Bill of Rights; it also assumed that individuals, regardless of which race to which they "belonged," would rise to the top. But was this really so? Here is where the issue was joined between the two perspectives. That American credo, as Myrdal dubbed it, was a problem, too, for its claimants stressed the all-sufficiency of the meritocratic individual. Yet in turn such was problematic, for that individualism was a credo of conformity to white middle-class group standards of conduct. Myrdal's idealism, itself a product of his antifascist attitudes, helped "sell" his argument to postwar social scientists and public policymakers. Clearly only well-mannered middle-class blacks would become integrated with the white middle class. The (presumably) ruder black masses could never become meritocratic individuals. In other words, white racism was still alive and well in postwar American middle-class culture. It had just become tacit, subterranean.

Hence the reaction to Myrdal's study is important for us to understand, for it helped convince the white elite classes of a certain approach to "the Negro problem" and in ways that flattered those in these elite classes. Harvard graduates no less than Old Blues from New Haven could believe that they were where they were solely because of their sterling qualities as individual persons and professionals. As much of the social science literature of the professions in the generation or so after World War II easily demonstrates, such as Harvard sociologist Talcott Parsons's *The Structure of Social Action* (1948), competition and status in that affluent American postwar society were matters of individual merit, and criticism of this individualistic argument was usually dismissed by academics and social scientists as either Marxian propaganda, not a helpful label in the heyday of the Cold War, or as bombastic silliness, such as Theodore Caplow and Reece J. McGee's hilarious but trenchant *The*

Academic Marketplace (1958), or even Kingsley Amis's wonderfully witty novel *Lucky Jim* (1960), a tale of academic misadventures and nonmeritocratic situations and factors in academic success if there ever were one.[35] The constitutional-political-cum-individual-merit-argument that Myrdal had done so much to anoint and to disseminate beyond the academy seemed to carry the day in postwar America—right down to 1965. One even found civil rights leaders, even black leaders who presumably understood the dynamics of the black community, using it, if for no other reason than it was the coin of the realm in public policy on the issue of race in American life then.[36] It was the discourse's sun, so bright and beguiling that it virtually eclipsed the discourse's moon, that of social problems and victimization. If the sun and the moon of this discourse still belonged to the same solar system, it was nevertheless the sun that most Americans, especially white Americans, came to notice. America was still a white man's country.[37]

The End of Confederate Historiography

Yet major changes within the academy and national politics were in the wind in postwar America. Among scholars, the important shift was a radical new reinterpretation of American slavery and its significance in American history, past and present. In the interwar years it was Ulrich B. Phillips, a historian at the Universities of Michigan and Yale, who set the parameters of the scholarly interpretation of American slavery. Phillips was the first historian to use, with any claim to rigor or comprehensiveness, archival records to depict slavery as an institution; in *American Negro Slavery* (1918) and, more elegantly, in *Life and Labor in the Old South* (1929), Phillips portrayed slavery as an essentially benign institution that created a web of relationships between masters and their slaves—economic, social, cultural, legal, and the like—that was essentially beneficial for all concerned. It was what could be dubbed a Confederate interpretation, which was hardly surprising, for Phillips himself was a son of the Old South, which he saw through rose-tinted, racist glasses. Nothing could have conflicted more with Frazier's interpretation of American slavery and its role in American history, past and present, than Phillips's view; but Phillips's view prevailed over Frazier's, even to the extent that it was enshrined in widely adopted American history college texts, such as Samuel Eliot Morison and Henry Steele Commager's *The Growth of the American Republic* (1942).[38]

If there could be "Confederate" views of the history and significance of American slavery, there could also be "abolitionist" ones as well.[39] In *Slave and Citizen: The Negro in the Americas* (1947), Frank Tannenbaum, a professor of Latin American history at Columbia University, addressed the problem of slavery in the Western Hemisphere as the problem of Africans in the New World and what happened to them. He drew a sharp contrast between the experiences of Africans in Latin America and in North America, showing that the laws and customs of slavery in the colonies of Spain and Portugal provided for the protection of slaves against cruelty, and even ways of self-emancipation, whereas in the United States, given that this was chattel slavery, African Americans were trapped in a racial box: slavery. Tannenbaum wrote the book under the influence of World War II; he argued that the meaning of Western culture was to protect the moral equality of the individual—an obvious reference to Nazi Germany, in that context—and that the history of slavery in the Americas showed precisely that point.[40]

In the 1950s two young American historians, Kenneth M. Stampp and Stanley Elkins, effected a radical change in the picture that most American historians had of the role and significance of American slavery in American history. Both were aware of the radical environmentalism that such experts in theories of culture and personality as John Dollard and Robert R. Sears of Yale, Kurt Lewin of Iowa, Margaret Mead of the American Museum of Natural History, and Gordon Allport of Harvard had used in their own work.[41] Some of this work, especially by Holocaust survivors such as Bruno Bettelheim, stressed that in the German concentration camps—in extreme and total environments, that is—the inmates could regress in their behavior to near-infantile status. Institutions, in short, could radically warp personality and culture—a chilling prospect indeed.[42] In *The Peculiar Institution: Slavery in the Ante-Bellum South* (1956), Stampp took Phillips's work to task; indeed, in many respects his book was a point-by-point refutation of *Life and Labor in the Old South.* Stampp was more aware of the behavioral and social science literature on personality and culture than he was often given credit for, but his concern in the book was to present a solidly researched historical alternative to Phillips's scholarship. And most historians greeted his work enthusiastically unless they had prosouthern sympathies. In particular, black historians were pleased, for they had chafed at Phillips's apologia for slavery and insisted on slavery's cruelty, harshness, and brutality.[43] Stampp's racial egalitarianism originated with the maxim that race was only skin deep, that black people were just white people with black skins. "Today we are learning much from the natural and social sciences about the Negro's potentialities and about the basic irrelevance of race, and we are slowly discovering the roots and meaning of human behavior," Stampp insisted. He discussed slavery without the sugar-coated bromides that Phillips had deployed; for example, when he took up the matter of whites murdering slaves, he wrote that "the great majority of whites who . . . were guilty of feloniously killing slaves escaped without any punishment at all."[44] Even today it is difficult to read Stampp's book without a sense of horror and dismay.

In *Slavery: A Problem in American Institutional and Intellectual Life* (1959), Elkins, who was far more self-consciously interdisciplinary than Stampp (or the majority of American historians), took the same psychological and anthropological racial environmentalism as his point of departure as had Stampp, but he came to different conclusions. He welded Tannenbaum's insights about the comparisons—contrasts, really—between the slave regimes in the United States and Latin America together with the contributions of the personality and culture theorists, especially those who argued that such total institutions as concentration camps could radically distort the personalities of their inmates, and fashioned an original, controversial, and chilling interpretation of North American slavery that seemed to show how slavery warped black institutions, including the black family, and fanned the fires of institutionalized white racism against blacks. Thus the full indictment of slavery and racism that African American scholars such as Frazier had urged for so many years and which had been ignored by the larger white society, was now an established, if somewhat controversial, interpretation in the mainstream of the American historical profession. If many professional historians criticized his book when they reviewed it for scholarly journals, nevertheless it gained a certain currency. Especially in the early to mid–1960s, it gained quite a following among teachers, who saw the book as stimulating reading, no matter what their opinion of it might be, and some historians also found considerable value in it as a bold and suggestive guide to some of the more difficult and contentious problems in the history of slavery and race relations. And Elkins won support outside the academy.[45] Thus Charles E. Silberman, in his widely read journalistic account, *Crisis in Black and White* (1964), gave Elkins a rave review for suggesting environmental (and therefore nonracist) explanations for certain "unpleasant facts" about African Americans that white liberals and radicals did not wish to concede to white racists, such as promiscuity, crime, family stability, and IQ test scores, where blacks did appear as a group to do less well than whites. Elkins had written "the most brilliant and probing study of slavery in the United States," cooed Silberman, and Silberman certainly provided an articulate and cogent thumbnail sketch of Elkins's arguments that a white liberal or radical could easily accept.[46]

By the time Elkins published his book, events in the political arena had brought on a national political crisis in what was then known as the civil rights problem. Black militancy had grown after World War II, and as African Americans gained allies among white liberals and radicals within and without the two major political parties, they began to become more self-confident about their prospects for a better life in the future.[47] On May 17, 1954, the U.S. Supreme Court handed down a decision that altered the very context of this history, because for the first time African Americans were no longer mere passive objects of the white mainstream political system. That decision made American blacks into active subjects and therefore participants in the group struggles and competitions of the national political system. The Court's justices unanimously ruled in *Brown v. Board of Education of Topeka*, that segregation of public schools, as sanctioned by its 1896 decision, *Plessy v. Ferguson*, was unconstitutional. Chief Justice Earl Warren wrote the decision for the Court. Very carefully he disentangled the decision on public education from the alleged central issue of *Plessy*, transportation, so that *Plessy* was not overturned. Thurgood Marshall and his colleagues on the National Association for the Advancement of Colored People's legal team who sued for the plaintiffs were afraid to ask the Court to reverse itself on *Plessy* for tactical reasons, and Warren, who had served only two months as chief justice when he penned the decision, believed, as did the other justices, that overturning *Plessy* would needlessly inflame the country to no good end.[48]

A close reading of *Plessy* and of *Brown* underlines the basic point that these decisions were about groups of people—races, that is—and had nothing to do with individuals in any way whatsoever. The heart of *Plessy* was simple and easy to understand; it turned on the constitutionality of a law in Louisiana that permitted the segregation of the white and colored races on railroad passenger cars. The justices ruled that not only was the plaintiff, who happened to be of mixed white and black ancestry, a black man, but that the law itself was sustained because many states could and did rule variously about the relations of the white and nonwhite races, and these state laws were beyond the protection of the federal Constitution, especially the Thirteenth and Fourteenth Amendments. The federalism that the justices contemplated in 1896 assumed that, short of the actual violation of the letter of the federal Constitution by a state law, the states could establish any regulations they wished regarding racial relations.[49]

In 1954 Warren and his colleagues virtually ignored any such intricate schemes of federal-state relations. They declared that much had changed since *Plessy*; public education was now an important activity of the states and municipalities, and it had no relevance whatsoever to transportation. Because there was considerable historical and social scientific evidence that segregation harmed African American children and made their prospects for a decent life as citizens bleak or worse, segregation of public schools was unconstitutional because it violated the Fourteenth Amendment's equal protection clause. The physical facilities of the white and black schools in the cases at hand might well have been "equal," but the psychological influences of each situation massively favored white over African American children. Although Warren did not cite the work of Frazier, the Frazier-Elkins hypothesis that slavery had made all but permanent victims of African Americans was definitely the undergirding of his argument, and thus the argument was injected into the nation's public political, educational, and cultural discourses concerning racial matters.[50]

Nor was this all. If the first *Brown* decision, known among constitutional experts as *Brown I*, declared blacks to be a group in the body politic, the Court's implementation decision the next year, or *Brown II*, suddenly transformed them back to individual petitioners, thus fudging the question constitutionally as to whether they were a legitimate group in the political system as this idea is understood in this chapter.[51] No wonder there has been confusion in the public discourse over these matters—and enormous difficulty for black Americans to gain their rights and prospects as citizens.

It was perhaps no accident that as a rising young social scientist Daniel Patrick Moynihan was introduced to the Frazier-cum-Elkins argument by no less a bright star in contemporary social science than Nathan Glazer, with whom he cooperated on several projects before entering government service in the Kennedy administration.[52] Of course, by the time Moynihan published his report on the African American family, which in turn led to President Johnson's speech at Howard University and the administration's official launching of affirmative action as a major object of public policy, scholarly opinion had gone beyond the Frazier-Elkins-Moynihan argument about victimization of the black family leading to all manner of social problems in the African American community.[53] And where did that color-blind ideology of individualism which our friends, the conservative constitutional historians, celebrate come from? Clearly it was a part of the ideology of the new professional middle classes of postwar America, including especially the sons and grandsons of immigrants from southern and eastern Europe. And it was also part and parcel of the rhetoric and ideology of the Southern Christian Leadership Conference as led by Martin Luther King Jr. and his associates. In no important way did it relate to the issues of individual behavior; rather, it was a statement that various groups could now struggle in the arena of mainstream American politics, citing all the while one of the hoariest of all American political mythologies.[54] But such ironies and paradoxes are indeed endemic in the worlds of social science and public policy.

Notes

I would like to thank Mary O. Furner, University of California, Santa Barbara; David M. Katzman, University of Kansas; Henry D. Shapiro, University of Cincinnati; and Clarence E. Walker, University of California, Davis, for commenting on an earlier draft of this chapter.

1. Michael Harrington, The *Other America: Poverty in the United States* (New York: Macmillan, 1962).
2. Noel Ignatiev, *How the Irish Became White* (New York: Routledge, 1995), and Matthew Frye Jacobson, *Whiteness of a Different Color: European Immigrants and the Alchemy of Race* (Cambridge: Harvard University Press, 1998) are good examples of this literature. One need not accept all their assumptions and arguments to see the essential point these writers are making: that whiteness and group identity were functionally related in the American political system.
3. On Johnson, see, for example, Robert A. Caro, *The Years of Lyndon Johnson: The Path to Power* (New York: Alfred A. Knopf, 1982); Robert A. Caro, *Means of Ascent: The Years of Lyndon B. Johnson* (New York: Alfred A. Knopf, 1990); Robert Dallek, *Lone Star Rising: Lyndon Johnson and His Times* (New York: Oxford University Press, 1991); Robert Dallek, *Flawed Giant: Lyndon Johnson, 1960–1973* (New York: Oxford University Press, 1998).
4. Lee Rainwater and William L. Yancey, eds., *The Moynihan Report and the Politics of Controversy*, A Trans-Action Social Science and Public Policy Report (Cambridge: MIT Press, 1967), reprints both the Howard University speech and Moynihan's pamphlet.
5. Paul Moreno, *From Direct Action to Affirmative Action: Fair Employment Law and Policy in America, 1933–1972* (Baton Rouge: Louisiana State University Press, 1997), 1–29; Herman Belz, *Equality Transformed: A Quarter Century of Affirmative Action* (New Brunswick, N.J.: Transaction Publishers and the Social Philosophy and Policy Center, 1991), 1–42.
6. John Cawelti, *Apostles of the Self-Made Man* (Chicago: University of Chicago Press, 1965), makes this point well.
7. Thomas L. Haskell, *The Emergence of Professional Social Science: The American Social Science Association and the Nineteenth Century Crisis of Authority* (Urbana: University of Illinois Press, 1977).
8. This is evident in the secondary literature of the democratic revolution of the 1830s and 1840s: see Hamilton Cravens, "History of the Social Sciences," in *Historical Writing on American Science: Perspectives and Prospects*, ed. Sally G. Kolhstedt and Margaret Rossiter (Baltimore: Johns Hopkins University Press, 1986), 183–207; and Hamilton Cravens, Alan I Marcus, and David M. Katzman, eds., *Technical Knowledge in American Culture: Science, Technology, and Medicine Since the Early 1800s* (Tuscaloosa: University of Alabama Press, 1996).
9. This is based on my work for a book in progress on the history of the social and behavioral sciences in America, "Designing Humanity: The Social and Behavioral Sciences in American Culture since the Revolution."
10. See, for example, Cawelti, *Apostles*; Richard White, *The Middle Ground: Indians, Empires and Republics in the Great Lakes Region 1650–1815* (New York: Cambridge University Press, 1991).
11. R. D. McKenzie, *The Metropolitan Community* (1933; New York: Russell and Russell, 1967), 69.
12. Robert E. Park, foreword to *The Ghetto*, by Louis Wirth (Chicago: University of Chicago Press, 1928), viii–ix.

13. Roscoe C. Hinkle Jr. and Gisela J. Hinkle, *The Development of Modern Sociology: Its Nature and Growth in the United States* (New York: Random House, 1954), 18–43, esp. 28–37.
14. It may be presumed that Franz Boas, the fiercely antiracist Columbia University anthropologist, was at least somewhat influential in persuading Thomas to take up such views; see Franz Boas, *The Mind of Primitive Man* (New York: Macmillan, 1911). See also George W. Stocking Jr., *Race, Culture, and Evolution: Essays in the History of Anthropology* (New York: Free Press, 1968); Hamilton Cravens, *The Triumph of Evolution: The Heredity-Environment Controversy, 1900–1941* (Baltimore: Johns Hopkins University Press, 1988 [1978]), chs. 4, 5.
15. William I. Thomas and Florian Znanecki, *The Polish Peasant in Europe and America*, 5 vols. (Chicago: University of Chicago Press, 1918).
16. See Cravens, "History of the Social Sciences." See also Cravens, *Triumph of Evolution*; and Hamilton Cravens, *Before Head Start: The Iowa Station and America's Children* (Chapel Hill: University of North Carolina Press, 1993). I am developing this argument— the centrality of the group and the unimportance of the individual in all social science theory and discourse of whatever kind—in "Designing Humanity."
17. See Stow Persons's acute *Ethnic Studies at Chicago 1905–45* (Urbana: University of Illinois Press, 1987), esp. 28–98. An institutional history of the Chicago school that is quite useful is Martin Bulmer, *The Chicago School of Sociology* (New York: Cambridge University Press, 1985).
18. Wirth, *The Ghetto*, 193, 281.
19. Cravens, *The Triumph of Evolution*, 157–274; Mary O. Furner, *Advocacy and Objectivity: A Crisis in the Professionalization of American Social Science, 1865–1905* (Lexington: University Press of Kentucky, 1975).
20. On Park and Reuter, I have depended heavily upon Persons, *Ethnic Studies at Chicago*, 60–130.
21. E. Franklin Frazier, *The Negro Family in Chicago* (Chicago: University of Chicago Press, 1932); E. Franklin Frazier, *The Free Negro Family: A Study of Family Origins before the Civil War* (Nashville: Fisk University Press, 1932), 72.
22. E. Franklin Frazier, *The Negro in the United States* (Chicago: University of Chicago Press, 1939); Persons, *Ethnic Studies at Chicago*, 131–150; Dale R. Vlasek, "E. Franklin Frazier and the Problem of Assimilation," in *Ideas in America's Culture: From Republic to Mass Society*, ed. Hamilton Cravens (Ames: Iowa State University Press, 1982), 141–155.
23. The four studies were: Allison Davis and John Dollard, *Children of Bondage: The Personality Development of Negro Youth in the Urban South* (Washington, D.C.: American Council on Education, 1941); E. Franklin Frazier, *Negro Youth at the Crossroads: Their Personality Development in the Middle States* (Washington, D.C.: American Council on Education, 1940); Charles S. Johnson, *Growing Up in the Black Belt: Negro Youth in the Rural South* (Washington, D.C.: American Council on Education, 1940); W. Lloyd Warner, Buford H. Junker, and Walter A. Adams, *Color and Human Nature: Negro Personality Development in a Northern City* (Washington, D.C.: American Council on Education, 1941).
24. Frazier, in *The Negro Family in the United States*, advanced the victimization argument in full battle dress.
25. Frazier, *Negro Youth at the Crossroads*, 29.
26. Warner, Junker, and Adams, *Color and Human Nature*, 295, 296.
27. Davis and Dollard, *Children of Bondage*; Johnson, *Growing Up in the Black Belt*.

28. Ira De A. Reid, *In a Minor Key: Negro Youth in Story and Fact* (Washington, D.C.: American Council on Education, 1940), 21, 37; calculations mine.
29. Allison Davis, Burleigh B. Gardner, and Mary R. Gardner, directed by W. Lloyd Warner, *Deep South: A Social Anthropological Study of Caste and Class* (Chicago: University of Chicago Press, 1941).
30. W. Allison Davis and Robert J. Havighurst, *Father of the Man: How Your Child Gets His Personality* (Boston: Houghton Mifflin, 1947); Kenneth Eells, Allison Davis, Robert J. Havighurst, Virgil E. Herrick, Ralph W. Tyler, under the chairmanship of Allison Davis, *Intelligence and Cultural Differences: A Study of Cultural Learning and Problem-Solving* (Chicago: University of Chicago Press, 1951); quote in Allison Davis, *Social-Class Influences Upon Learning, The Inglis Lecture 1948* (Cambridge: Harvard University Press, 1948), 90.
31. For a useful history of this study, see Walter A. Jackson, *Gunnar Myrdal and America's Conscience: Social Engineering and Racial Liberalism, 1938–1987* (Chapel Hill: University of North Carolina Press, 1990).
32. The works that came directly through the Carnegie Corporation's sponsorship as studies that accompanied Myrdal's book were: Melville J. Herskovits, *The Myth of the Negro Past* (New York: Harper and Row, 1941); Charles S. Johnson, *Patterns of Negro Segregation* (New York: Harper and Row, 1943); Richard Sterner, *The Negro's Share* (New York: Harper and Brothers, 1944); Otto Klineberg, ed., *Characteristics of the American Negro* (New York: Harper and Brothers, 1944).
33. For a full discussion of this work in the human sciences, see Cravens, *The Triumph of Evolution*, ch. 5.
34. Gunnar Myrdal, with the assistance of Richard Sterner and Arnold Rose, *An American Dilemma: The Negro Problem and Modern Democracy* (New York: Harper and Brothers, 1944), 997, 1004, 1021, and 1024.
35. Talcott Parsons, *The Structure of Social Action*, 2nd ed. (1937; Glencoe, Ill.: Free Press, 1948), was much more influential in its second than in its first edition because its individualistic point of view, which came from Parsons's background as a laissez-faire classical economist before becoming a sociologist, was much more popular in the 1950s than in the 1930s and 1940s. Theodore Caplow and Reece J. McGee, *The Academic Marketplace* (New York: Basic Books, 1958), with an introduction by Jacques Barzun; Kingsley Amis, *Lucky Jim* (New York: Viking Press, 1960).
36. See, for example, President's Committee on Civil Rights, *To Secure These Rights* (Washington, D.C.: Government Printing Office, 1947); August Meier and Elliott Rudwick, *From Plantation to Ghetto*, rev. ed. (1965; New York: Hill and Wang, 1970), 251–298; Harvard Sitkoff, *The Struggle for Black Equality, 1954–1980* (New York: Hill and Wang, 1981); Jackson, *Gunnar Myrdal*, ch. 7; Richard Kluger, *Simple Justice: The History of Brown v. Board of Education and Black America's Struggle for Equality* (New York: Alfred A. Knopf, 1975), 700–778.
37. Jackson, *Gunnar Myrdal*, chs. 5–8, covers in considerable detail the implications of Myrdal's message and how widely and broadly it was received.
38. Ulrich B. Phillips, *American Negro Slavery: A Survey of the Supply, Employment, and Control of Negro Slavery as Determined by the Plantation Regime* (1918; New York: P. Smith, 1952); Ulrich B. Phillips, *Life and Labor in the Old South* (Boston: Little, Brown, 1929); Samuel Eliot Morison and Henry Steele Commager, *The Growth of the American Republic*, 2 vols. (New York: Oxford University Press, 1942); in general, see August Meier and Elliot Rudwick, *Black History and the Historical Profession 1915–1980* (Urbana: University of Illinois Press, 1986).
39. Thomas J. Pressly, *Americans Interpret Their Civil War* (Princeton, N.J.: Princeton

University Press, 1949), provocatively argues that all the modern scholarly interpretations of the Civil War were first expressed by participants and contemporaries in the Civil War era.

40. Frank Tannenbaum, *Slave and Citizen: The Negro in the Americas* (New York: Alfred A. Knopf, 1947), 93, vii–ix.
41. John Dollard, Neal E. Miller, Leonard W. Doob, O. H. Mowrer, and Robert R. Sears, *Frustration and Aggression* (New Haven, Conn.: Published for the Institute of Human Relations by Yale University Press, 1939); Roger Barker, Tamara Dembo, and Kurt Lewin, *Frustration and Regression: An Experiment with Young Children*. Studies in Topological and Vector Psychology II, University of Iowa Studies, Studies in Child Welfare, vol. 18, no. 1 (Iowa City: University of Iowa Press, 1941); Ronald Lippitt, *An Experimental Study of the Effect of Democratic and Authoritarian Group Structures*. Studies in Topological and Vector Psychology, II, University of Iowa Studies, Studies in Child Welfare, vol. 18, no. 1 (Iowa City: University of Iowa Press, 1940); Kurt Lewin and Ronald Lippitt, "An Experimental Approach to the Study of Autocracy and Democracy: A Preliminary Note," *Sociometry* 1 (1938): 292–300; Kurt Lewin, "Field Theory and Experiment in Social Psychology: Autocratic and Democratic Group Atmospheres," *American Journal of Sociology* 45 (1939): 26–49; Cravens, *Before Head Start*, 157–169, 188–191, 217–250; Gordon Allport, *The Nature of Prejudice* (New York: Macmillan, 1948).
42. Stanley M. Elkins, *Slavery: A Problem in American Institutional and Intellectual Life* (1959; New York: Grossett and Dunlap, 1963), 81–140, discusses this literature, including Bettleheim.
43. Meier and Rudwick, *Black History*, 245–247, covers the reaction to Stampp's *The Peculiar Institution* among white and black historians; indeed, their book should be followed carefully, for it is a bonanza of rigorously researched information, and I have relied upon it at many points in this essay.
44. Kenneth M. Stampp, *The Peculiar Institution: Slavery in the Ante-Bellum South* (New York: Alfred A. Knopf, 1956), vii, 222.
45. Meier and Rudwick, *Black History*, 247–251.
46. Charles Silberman, *Crisis in Black and White* (New York: Vintage, 1964), 74–75.
47. James T. Patterson, *Grand Expectations: The United States 1945–1974* (New York: Oxford University Press, 1996), 375–406, is but one general survey of the period that cites a handful of key secondary accounts about the developing civil rights issue in the 1950s. The literature on that problem is, in fact, immense.
48. *Brown v. Board of Education of Topeka* 347 US 483 (1954); *Plessy v. Ferguson* 163 US 537 (1896); Kluger, *Simple Justice*, 700–710.
49. *Plessy v. Ferguson*, esp. 540–552.
50. *Brown v. Board of Education of Topeka*, esp. 486–496. Warren did cite Myrdal, *An American Dilemma*, as well as the research of black psychologist Kenneth Clark, whose work with dolls and young black children suggested that black children in segregated schools had serious self-esteem problems that hampered their ability to compete in the larger world—hence the victimization thesis.
51. *Brown et al. v. Board of Education of Topeka* 349 U.S. 294 (1955) is the *Brown II* decision. See James T. Patterson, *Brown v. Board of Education: A Civil Rights Milestone and Its Troubled Legacy* (New York: Oxford University Press, 2001), in my judgment a rather too sanguine argument. On this point, see in particular Lewis M. Steel, "Separate and Unequal, By Design," *Nation* 272 (February 5, 2001): 27–32.
52. Nathan Glazer and Daniel Patrick Moynihan, *Beyond the Melting Pot: The Negroes, Puerto Ricans, Jews, Italians, and Irish of New York City* (Cambridge: MIT Press,

1963), whose fundamental thesis was that there was no such thing as a melting pot: all groups remained apart from all others. Glazer, of course, wrote the introduction to the paperback edition of Elkins's book; see Elkins, *Slavery*, ix–xvi.

53. Rainwater and Yancey, *The Moynihan Report*; see also Meier and Rudwick, *Black History*, 246–276.
54. A provocative treatment of what happened to civil rights is Thomas Byrne Edsall, with Mary D. Edsall, *Chain Reaction: The Impact of Race, Rights, and Taxes on American Politics* (New York: W. W. Norton, 1991).

Chapter 2

Statecraft and Its Retainers

American Economics and Public Purpose after Depression and War

MICHAEL A. BERNSTEIN

NOT LEAST OF THE historical forces that shaped the continuing evolution of the American economics profession in the latter half of the twentieth century was the unique prosperity the nation enjoyed throughout the 1950s and 1960s. If the application of a new learning to the management of a "mixed economy" provided an exceptional opportunity for social scientific expertise to demonstrate its rigor and effectiveness, the context within which that display took place set the terms of both its practice and its success. Having proved its mettle in the extraordinary years of world war and having continued to do so in the early stages of what would be an even longer Cold War, modern economic theory was now deployed in an altogether novel exercise—the pursuit and maintenance of full employment growth in peacetime. That, owing to history itself, the national economy was singularly well positioned for sustained expansion in the postwar period made that task all the more tractable.

Unlike any other industrialized nation in the world at the time, the United States met the 1950s with an economy not only physically intact but organizationally and technologically robust. The demographic echoes of war set the stage for an acceleration in the rate of population growth, while the labor market effects of demobilization surprisingly sparked a rise in wages and incomes. Rapid and profitable conversion to domestic production was further engrossed by foreign demand—most vividly and poignantly emanating from those regions most devastated by the war itself—for the products of American industry and agriculture. As for international finance, the nation stood as creditor virtually to the entire world, and the dollar, both by default and by a multilateral agreement first reached by the Allied nations at Bretton Woods, had become a kind of *numeraire* to a newly emergent system of global commerce. With no small justification, the 1950s and 1960s would come to be regarded as a "golden age" of American capitalism.[1]

Macroeconomic management, demanding under any circumstances, was made substantially easier for a postwar generation that found itself the beneficiaries of historical circumstance. Far from solving the cruel puzzle of idle capacity and widespread unemployment that had characterized the Great Depression and unlike the challenge to rationalize allocation and maximize production in the emergency of war, the task that lay before American economists by the mid–1950s was both more straightforward and less difficult. More straightforward because, thanks to both the "Keynesian revolution" in economic thought and the policy experience derived from mobilization and war, the relationship between individual market behavior and aggregate outcomes was finally subject to systematic understanding. Less difficult because, given the sturdy rebound of the economy in the wake of World War II, there existed both the confidence (most especially exemplified by the moderate rates of return in the markets for Treasury bills and other government obligations) and the means (most vividly represented by rising income tax receipts) to realize fiscal spending targets with a minimum of redistributive implications.

So optimistic were politicians and the vast majority of economists concerning the effectiveness of stabilization policy techniques that it became fashionable by the early 1960s to speak of the "end of the business cycle" and of the ability of policy-makers to "fine tune" macroeconomic performance. In the 1965 report of his Council of Economic Advisers (CEA), President Lyndon Johnson made it clear that he "d[id] not believe recessions [we]re inevitable." Similarly, in what was arguably the most influential economics textbook ever published, Paul Samuelson wrote that his colleagues "*kn[ew] how* to use monetary and fiscal policy to keep any recessions that br[oke] out from snowballing into *lasting* chronic slumps." Expert knowledge buttressed by a healthy and resilient economy could now make the periodic deprivation and hardship once believed to be the inevitable consequence of the cycle truly a thing of the past.[2]

A REMARKABLY prosperous decade in the United States, the 1950s were nevertheless punctuated by three recessions. Relatively brief and mild, these downturns stood as a sturdy challenge to mainstream macroeconomists who believed that a new learning could make such fluctuations a thing of the past.[3] Such downturns also assumed, especially in the case of the last slump (which occurred on the eve of the 1960 presidential campaign), a growing significance in the minds of politicians eager to "score points" in electoral contests that had been, at least since the thirty-fourth president's reelection in 1956, fairly tame. For Massachusetts Senator John F. Kennedy, in the very closely contested race for the presidency in 1960, tarring his opponent, Vice President Richard Nixon, with the brush of the 1959 recession was a useful and ultimately successful, if decidedly opportunistic, tactic.[4]

By the winter of 1961, therefore, a sluggish national economy had become both a political liability for a new administration and a professional opportunity for a new cadre of economists in the Executive Office Building. On the one side, a new president faced a test that would, among others, ultimately determine if he might join Franklin Roosevelt and Woodrow Wilson as the only Democrats since Andrew Jackson to serve more than one consecutive term in the White House. On the other, Kennedy's experts had the chance to show what a "New Economics" could do; indeed, they might at last demonstrate, absent the distortions of depression and world war, of what a modern industrial (and well-managed) economy was capable in peacetime. To this heady set of circumstances were brought the striking sophistication on economic matters of a new president and the powerful skills of analysis and persuasion of his handpicked CEA. For the economics profession as a whole, it was an exquisite moment of anticipation and confidence.

On the secure professional foundation bequeathed to them by their immediate predecessors, the economists of Kennedy's New Frontier fashioned a spirit of unabashed activism.[5] Acknowledging that "policy-making [wa]s by no means an exact and objective science," Yale's James Tobin, who joined the CEA just after Kennedy's inauguration, firmly believed that a "neutral nonpartisan Council, if one could be imagined, would simply not provide advice of interest to the President." The Chief Executive, in his view, surely needed "professional" as well as "disinterested advice," but it necessarily had to come from those who "share[d] his objectives and his concern for the record of his Administration." Even more to the point, to the extent "economics ha[d] always been a policy-oriented subject," its application to "the urgent . . . issues of the day" was both appropriate and essential. Without it, the discipline "w[ould] become a sterile exercise, without use or interest." Not surprisingly, and in a fashion his council colleagues Kermit Gordon and Walter Heller would have heartily endorsed, Tobin had no patience with those "who fear[ed] that economics w[ould] be discredited if it [wa]s applied" to contemporary debates; they "reminded [him] of a football coach who never play[ed] his star back for fear he might be injured."[6]

To a not insignificant degree, the pointed determination of Kennedy's advisers to locate themselves at the center of economic policy disputes had as much to do with personality and temperament as with particular convictions regarding professional obligation and opportunity. Gordon, for example, was a skilled administrator endowed with a fine sense of political trends and a wise perception of the president's concerns on any given topic. The CEA chair, Walter Heller, had a remarkable gift for writing brief, insightful, and thoroughly convincing memorandums that not only captured Kennedy's attention for the CEA's work but also, over time, encouraged the president in the practice of turning to the Council ever more frequently for analysis and consultation. As one distinguished historian of the period put it, Heller's "pithy prose . . . carried a force in a page or two that the [Department of the] Treasury's tedious, bureaucratic reports could not match." Finally, while somewhat shy and diffident, Tobin possessed a powerful intellect that served, for his more gregarious and publicly visible colleagues, to ground the Council's work on the sturdiest possible analytic grounds. A remarkable team, Gordon, Heller, and Tobin have an exceedingly strong claim to have been the most impressive and certainly the best-known CEA ever to have served the Chief Executive.[7]

Yet above and beyond the fine synergy that animated their days in Washington, Kennedy's Council members also shared commonalities of experience and training of a more systematic sort. All three had begun their graduate work during the Great Depression, forced by circumstance thereby to bring to their introduction to the advanced study of economics a worldliness that made them impatient with abstraction for its own sake. At the same time, their own teachers and mentors were beginning to digest the withering attacks on orthodox theory fostered by the work of Keynes and his students, ultimately placing Gordon, Heller, and Tobin at the forefront of a new generation of Keynesian scholars in the United States. Perhaps even more important, like most of their generation, these men had been wrenched from their professional studies to serve the nation in World War II. In all three cases they became part of a vast apparatus of federal economic planning that had not only been born of necessity but also, most vividly, had worked well. It is hardly surprising that they would thus bring to Washington almost two decades later a faith in and commitment to macroeconomic management that was as determined as it was sophisticated.

Coupled with the talents and experience that made his CEA so effective were the patronage and support afforded by President Kennedy's receptiveness to economic argument. Indeed, it was this openness, a product of the "analytical mind" Kennedy brought to discussions of economic policy, that gave his CEA, especially in the person of Heller, unique access to the Oval Office. While formally the activism of the Kennedy Council may have seemed fairly similar to the widely criticized politicization witnessed during Leon Keyserling's tenure as chair, it had been the case that Harry Truman was far less aware of and interested in the work of the first CEA. The situation was altogether different in the winter of 1961 as the youngest president in the nation's history took office.[8]

Faced with an economy that had left the unemployment rate around 7 percent, the new administration in Washington was also discomfited by middling productivity gains in the nation's workplaces, which now weakened America's international trade position. What had been almost two decades of unchallenged national supremacy in world markets, a circumstance both facilitated and recognized by the Bretton Woods agreements of 1944, could no longer be sustained in the face of the revitalization of the economies of Western Europe and Japan. As they reestablished their international economic presence, nations like the Federal Republic of Germany and Japan exploited the advantages of an advanced technological base that was the outgrowth of the recent rebuilding of their major industries. Ironically enough, they also thrived because of their relative insulation, under international treaties and protocols (exemplified by the erection of a "nuclear umbrella" by the United States to forestall what was feared to be the potential for Soviet aggression), from the burdens of defense spending. Consequently, their major manufacturing sectors—such as automobiles, electronics, and steel—became powerful competitors with their American counterparts. Whatever the concerns of President Kennedy's CEA with the domestic weaknesses of the national economy, the international context within which these difficulties emerged could not be ignored.[9]

Given these fairly stark international realities, it was hardly surprising that some of the most powerful policy-makers in the Kennedy administration sought to frame the nation's economic challenges with respect to global financial networks. Both Treasury Secretary Douglas Dillon and his undersecretary for monetary affairs, Robert Roosa, regarded the growing imbalance between imports and exports, and the potential drain on national gold stocks of which it warned, to be the defining economic policy problem of the New Frontier. In this assessment, they were joined by William McChesney Martin, chair of the Federal Reserve System Board of Governors. As a central banker, Martin was further troubled by the inflationary bias that any deterioration in the value of the dollar (and thus in its "buying power") would engender. Both Treasury and the "Fed" were thus of like mind that relatively high interest rates were, by late 1961 and early 1962, a desirable and appropriate goal of administration economic policy.[10]

For President Kennedy's CEA, however, no matter how customary and venerable the medicine, the proposed monetary cure was worse than the fiscal disease. If the productivity of enterprise could only be enhanced by new investment and if unemployment could only be reduced by expenditures large enough to expand capacity and output, raising the price of loanable funds would be both counterproductive and self-defeating. The solution lay, in the view of Heller and his colleagues, in stimulating economic growth through deficit spending while accommodating such "pump priming" through flexible interest rates.[11] In peacetime the nation has never witnessed a more vivid confrontation between the goals of full employment activism and traditional monetary propriety. Exemplified by the bureaucratic struggles that emerged between the CEA and the Fed, the debate over the proper "mix" of fiscal and monetary policy during the Kennedy administration would become emblematic of national policy discussions through the remainder of the century.

Late in 1961, the CEA members began to formulate a plan to bring unemployment down to the 4 percent level—a target they believed both feasible and necessary to bring the economy out of the doldrums. In their view, the most efficient and politically expedient method to reach that target was through an income tax cut. By thus increasing the amount of discretionary funds available to producers and consumers alike, the tax reduction would, they believed, stimulate spending that would ultimately generate higher levels of employment and national product. The short-run deficits incurred by the loss of federal revenue would be redeemed in the longer run by a larger tax yield (even at lower tax rates) from a higher national income. Managing demand in the nation's marketplace in this fashion would cause an increase in the supply of goods and services (and thereby of incomes across the board) that would enable the economy to grow out of the initial red ink.[12]

In proposing the first deliberate peacetime indulgence of federal budget deficits, Gordon, Heller, and Tobin faced a formidable task of both persuasion (of the president and the Congress) and bureaucratic neutralization (of the Fed and its allies in the Treasury). To a large extent, the former was more daunting than the latter. Convincing the president that a tax cut was appropriate actually rekindled a debate that had begun shortly after the Democratic Party convention in the summer of 1960. An annual macroeconomic growth target of 5 percent had been made part of the party's convention platform at Los Angeles; on behalf of the Democratic Advisory Committee, Leon Keyserling had prepared a manifesto that claimed, among other things, that realizing the goal required large increases in direct federal spending. It was not a position with which Kennedy was particularly comfortable, especially given the vigorous counsel he had received from his father regarding the virtues of tight money and balanced budgets. He turned to his economics "brain trust" to sort through the issues, thus setting the stage for a policy debate that continued through the first two years of his presidency.[13]

Sensing the Democratic nominee's concern "that a kind of unmitigated Keyserling or old-style Democratic liberalism in regard to economics and fiscal policy wasn't going to pay off politically . . . during the campaign," Tobin began formulating a tax reduction strategy to achieve the 5 percent growth target. While his arguments did not become part of Kennedy's campaign rhetoric, they did provide the foundation for discussions that would take place late in 1961 and early in 1962 within both the Executive Office Building and the Oval Office. Ironically enough, despite his agnosticism on the deficit question, during his campaign Kennedy did not avoid suggesting the need for an "easy money" policy as part of a growth package. The candidate's somewhat cavalier views on the matter caused a great deal of consternation at the Fed. As for the "fiscalists" in his inner circle, moving Kennedy beyond campaign speculation to presidential decision was another matter altogether.[14]

The president's relative degree of economic sophistication notwithstanding, it fell to his CEA advisers to make their case for a tax cut in a fashion that would both persuade and inspire. For this purpose, Heller's adroit skill in rendering policy argument as graceful prose linked up well with Tobin's sharply honed analytical instincts. Turning to CEA staff economist Arthur Okun, Tobin asked his former Yale colleague to estimate, if possible, the relationship between the level of unemployment and the magnitude of the gross national product. Out of that statistical protocol emerged "Okun's Law," a rather straightforward calculation that showed that for every 1 percent reduction in unemployment there could be garnered (through direct impacts on levels of output and indirect reductions in the "underemployment" of contracted labor in slack times) a 3 percent increase in national product. With that quantitative and rigorous demonstration of the virtues of a stimulatory fiscal policy, the Council captured Kennedy's imagination. Here was a powerful rhetorical device, one that Heller could easily exploit both in memorandums for the president and in testimony before the Joint Economic Committee of the Congress, with which to justify a tax cut and hold both budgetary conservatives and monetary purists at bay.[15]

Taking on congressional and public opposition to deficit spending was a political task in which the CEA became thoroughly engaged. The same can be said for the less visible in-fighting that emerged among traditionalists at both the Fed and the Treasury. In ways that presumably would have thrilled Leon Keyserling, the economists of the New Frontier did not hesitate to make clear their conviction that the reduction of unemployment, rather than of federal deficits, was the true mission of good government and the most compelling goal of a well-chosen public policy. Their ultimate success in making their case rested squarely on their skill in linking the analytical propositions of the New Economics with a thinly veiled ideological argument that the costs of idleness (both in foregone national income and social fragility) far outweighed the burdens of short-run debt and potential pressure on the international value of the dollar.[16]

In President Kennedy, Heller and his colleagues found a sympathetic student of the New Economics, nervous all the same about its political implications; in Fed chair Martin, and, to a lesser extent, Douglas Dillon at Treasury, they encountered more problematic skeptics. The timidity of his first budget message to the Congress notwithstanding, the president had refrained from asking for a tax increase to supplement additional military expenditures (between $3 billion and $4 billion) in the wake of the Berlin crisis.[17] What solace the CEA might have taken from that forbearance on the part of the Chief Executive was tempered by the knowledge of his oft-repeated fear that he "would be kicked in the balls by the opposition" if he asked outright for a tax cut. At the same time, as the president became more and more persuaded of the probity of the CEA's analysis of the nation's economic ills, "he suggested that the Council do some serious thinking about how to use the White House as a pulpit for public education." It was within this context that Tobin had turned to Okun for his landmark statistical study of unemployment and gross national product. As for Heller, he would later note that the profession had made "no greater contribution" than "raising the level of [economic policy] discussion" as had been done during the debate over the Kennedy tax reduction. Arthur Okun wholeheartedly concurred.[18]

Taking the measure of the naysayers at Treasury and the Federal Reserve Board of Governors had, by contrast, less to do with persuasive argumentation premised on scholarly credentials than with straightforward and hardheaded struggles for the president's ear. By far, Douglas Dillon was the easier opponent for the New Frontiersmen of the CEA. A lone Republican in a Democratic cabinet, his freedom of maneuvering was already quite constrained. More to the point, so profound was the mutual admiration shared between Heller and Undersecretary Roosa that the Kennedy Council enjoyed special access to the highest echelons in the Treasury Building. Here Heller's gift for writing, coupled with his gracious interpersonal style, truly paid off. Indeed, it was Dillon who had thrown crucial support to the CEA suggestion that the president avoid a tax increase for military purposes late in the summer of 1961.[19]

William McChesney Martin had neither the political obligations to President Kennedy nor the official responsibilities to the Executive Branch that constrained the conduct of Treasury Secretary Dillon. The "independence" of the Fed from the Executive Branch was the result of both conscious intent in its founding legislation and decades of practice among a board of governors whose sensibilities were more attuned to the needs of the nation's banking industry than anything else.[20] As a consequence, maintaining a conformity between fiscal and monetary policy was (and is) always difficult. Presidents have, more often than not, had to rely on the goodwill of Fed chairs to oblige their fiscal policy goals, while the Fed itself has had to prevail upon the willingness of particular administrations to refrain from tampering with its bureaucratic singularity. All these traditions and expectations the economics of the New Frontier worked to test.[21]

Martin, steeped in time-honored Fed practice and bearing the imprimatur of a Republican conservatism on monetary affairs belied by his appointment as chair by the Democrat Harry Truman, fully believed that the decision-making of the central monetary authority, and the appointment of its chief officers, should be "nonpolitical." It was for this reason that he had refused, contrary to the traditional script, to offer his resignation to the new president. That it was no less customary for the president to decline the offer made no difference, especially given rumors that Kennedy was more than ready to break with that habit. Martin, as early as January 1961, thus put Kennedy on notice of his intentions to buck the liberal tide in Washington. In Washington's activist climate in the early 1960s, the Fed chair's convictions were hardly immune from criticism. Even those predisposed to support Martin, given their earlier service to the Fed, found his obduracy inappropriate.[22]

Before he formally took office, Kennedy had anticipated difficulties with Martin and his colleagues. Indeed, the president elect had conceived of James Tobin as a prospective administration appointee who could "have [had] a real crack at some of th[e] policies of Bill Martin's." Yet whatever his intentions, Kennedy ultimately chose a road paved with compromise, negotiation, and close attention to his right flank in Congress; he ultimately reappointed Martin as Fed chair believing that he "need[ed] . . . Republicans [Martin and Douglas Dillon at Treasury] to maintain a strong front as far as the financial community [wa]s concerned." It was, however, more in the nature of a deal (however unacknowledged in public) than a surrender. By early 1963 the president encouraged his CEA to prepare, for inclusion in his 1963 budget message, the formal tax cut proposal so long debated and which he believed the Fed (in the person of its chair, now comforted by his renewed term and authority) would, if not endorse, simply tolerate. Its ultimate legacy was the Revenue Act of 1964. Peacetime deficit spending as an explicit growth policy of the federal government had finally come home.[23]

While most conspicuous among them, the CEA's struggles with the Fed were but part of a wider problem encountered by the Kennedy administration concerning the coordination of national economic policy. Here, too, Heller and his colleagues sought innovative and politically adept solutions. Born of their conviction that professional expertise had so much to offer the federal government and given their own schooling in the sometimes harsh context of interagency rivalry, these White House economists were eager to link the work of colleagues throughout the Executive Branch in ways that would avoid the misunderstandings and sometimes self-defeating actions of disparate offices. Out of their desire emerged an idea to forge structured links between the Council, the Treasury, and the Bureau of the Budget.

In what Walter Heller dubbed the "Troika," Kennedy administration officials from the CEA, Treasury, and Budget periodically met both to share information and interweave their respective agendas for policy formulation. The initial motivation for these meetings stemmed from the hope that Council forecasts of macroeconomic performance would be directly informed by annual revenue projections provided by Treasury and governmental expenditure estimates supplied from Budget. Over time, the sessions (held only when members had particular issues to place on a meeting docket) covered the entire range of policy questions facing the administration. After the struggles with Martin had been resolved, officials from the Fed were also included in an expanded format that was (less artfully) named the "Quadriad."[24]

To a large extent, and in ways that no doubt surprised its participants at least initially, the Quadriad arrangement worked well. Meetings went smoothly; operational matters were discussed with courtesy; and disagreements were resolved amicably. A bureaucratic counterpart to the detente that had emerged between the White House, the Treasury, and the Fed, the Quadriad served well through the balance of the Kennedy presidency and throughout the administration of Lyndon Johnson. Its very success masked a less obvious purpose, one even devious in design—the enhancement of the prestige and clout of the CEA itself. By placing the CEA on an equal footing with Treasury, Budget, and the Fed, the Quadriad setting officially lent the Council the kind of visibility and authority within the Executive Branch, akin to that of any cabinet office, that had been almost two decades in the making.[25] Indeed, this had been one of the concealed goals of Heller's gambit in the first place. Gardner Ackley would note years later that the Quadriad meetings were always conducted on terms set by the Council with the other original Troika members—Budget and Treasury—following its lead. As for the Fed, its participation was, in his view, "a joke."[26] Besting President Kennedy in the battle over his reappointment, Fed chair Martin had nonetheless lost the war.[27]

A little more than a year before he was murdered in Dallas, President Kennedy asked Congress to implement what would be the largest income tax reduction in the nation's history—a $13.5 billion decrease spread over three years. By early 1964 that proposal became law, coupled with a diminution in the levy imposed on corporate profits and the implementation of more generous depreciation allowances in the revenue code. Within a year the national unemployment rate fell to slightly over 4 percent, and the utilization of manufacturing capacity exceeded the 90 percent level in virtually all major sectors. By the fall of 1964 "the success of the tax cut" was so apparent that, in the words of Arthur Okun, "economists were riding about as high a crest of esteem and respect . . . as ha[d] ever been achieved." Perhaps it had been true, as Presidential Special Counsel Theodore Sorensen had told James Tobin in 1961, that "the most expendable thing" in the Kennedy White House had been the "reputations [of the CEA members] as professional economists." In the end, no matter how one valuated the stakes, the payoff to the gamble had been huge.[28]

If the apparent triumph of the New Economics brought immediate distinction to Heller and his Council colleagues, it also amassed honor for the profession as a whole. Across generational as well as political divides, economists from around the country shared in the accomplishments of their more visible colleagues in Washington "who brought the fruits of the Keynesian revolution firmly into the public field." Recent doctoral students could reflect with pleasure on the fact that the prestige of the CEA now brought "added luster to [the discipline] . . . and the[ir] degrees." Theoretical opponents, like Milton Friedman, hardly friends of the fiscal activism that had been practiced and now celebrated, could nonetheless take pride in the fact that they had anticipated that colleagues like Gordon, Heller, and Tobin would "acquit [their] responsibilities to the Administration & to the Profession . . . with integrity and competence." Similarly Arthur Burns believed that Heller could "leave [his] post with a feeling of achievement and in the knowledge that [he] ha[d] earned the respect of economists, even when they disagreed with [him.]" All members of the economics community had understood "that there [had been] a grave risk in the President surrounding himself with college professors," for "if they [had] failed they [would have] failed not only for themselves but for the group they represented." Yet their fears had been more than allayed. As a lead aide to Senator William Proxmire (Democrat of Wisconsin) had put it, before the Kennedy CEA had taken office "economics was viewed generally among top policy-makers, especially on Capitol Hill, as an esoteric field which could not bridge the gap to meet specific problems of concern." Heller and his associates "ha[d] almost single-handedly made the profession both respectable and useful in the eyes of government." Less than a century in the making, the professional community of American economists had every reason to be proud.[29]

So dramatic the apparent success of the New Economics and so intoxicating the faith and self-esteem it fostered among its proponents, it was perhaps inevitable that its architects would contemplate taking on even more demanding challenges. On the eve of President Kennedy's reelection campaign, Walter Heller, inspired by discussions with certain key advisers and armed with a striking examination of recent changes in income distribution undertaken by a CEA staff member, began to formulate a strategy to eradicate poverty nationwide. The president's receptivity to the idea had been in large part enhanced by the very positive impressions he had of the work of Michael Harrington, arguably the leading American socialist of his time. Yet it was not simply a concern with the problems of the poor that facilitated the acceptance of a new policy initiative on their behalf; the belief that modern macroeconomic management could both increase the growth rate and more equitably distribute its benefits was also no small part of the optimism that characterized the final days of Kennedy's presidency.[30]

A month before the president's fateful trip to Texas, the chair of his CEA began work on an antipoverty policy agenda. The links between the goals of eradicating want and achieving full employment output seemed to Walter Heller to be both obvious and straightforward. At the same time, a burgeoning civil rights movement and the pressures it brought to bear on a traditional Democratic Party strategy of tolerating southern segregation in exchange for electoral loyalty—a practice that had its roots in the secession crisis of the previous century—further encouraged an economic activism in Washington that was unprecedented. To most of the New Economists, the Kennedy tax cut had been "a necessary, although not sufficient, condition for the elimination of discrimination in employment" and the progressive redistribution of wealth and income to the nation's most unfortunate citizens. Having been "greatly troubled by the prospect of a civil rights program . . . launched without an expansionary economic program," they believed it a logical step to follow the fiscal triumphs of 1963 with a federal offensive against both deprivation and disempowerment. "[H]aving mounted a dramatic program for one disadvantaged group (the Negroes), it was [now] both equitable and politically attractive to launch [a] specifically designed [one] to aid other disadvantaged groups." Suddenly, far separated from abstruse debates about the relationship between fiscal and monetary policy, federal economists were now very much part of an effort to stimulate social and political change in modern American society that was as novel as it was complicated.[31]

No doubt inspired by the grief he nurtured during the long flight home from Love Field on the evening of November 22, 1963, Walter Heller went directly from Andrews Air Force Base to the Executive Office Building next door to the White House. He labored well into the night, preparing an extensive memorandum for President Lyndon Johnson that detailed both the current economic conditions facing the government and a blueprint for a national "War on Poverty." His efforts were requited the next evening when the president, taking him aside after a general staff meeting, urged him to "move ahead on the poverty theme . . . full-tilt." Declaring himself to be "a Roosevelt New Dealer," Johnson had further assured his chief economic adviser that he was "no budget slasher." He "underst[oo]d that expenditures ha[d] to keep on rising to keep pace with the population and help the economy." Fiscal and social activism had now become inextricably joined; their integument was the New Economics.[32]

Yet as the Berlin crisis had so tellingly demonstrated, any and all decisions regarding economic and social policy were necessarily framed during the 1960s within the context of the Cold War itself. Whatever domestic objectives he had set for himself, President Kennedy had continually been forced to reckon with the burdens imposed by a national security agenda that had emerged after 1945. The same was even more poignantly the case for his successor. In ways exquisitely ironic, the very defense apparatus that had provided such opportunity and resources to a social science that was regarded by the mid–1960s as "a great success" ultimately became a major part of its undoing. A civil war in the Indochinese peninsula in which the United States became embroiled for almost two decades was only the most vivid and costly example of that apostasy.

In the winter of 1961, CEA member Kermit Gordon received a letter from a colleague who believed that "the solution to the *economic* problems of Viet Nam . . . [wa]s fairly clear." Serving with the Military Assistance Advisory Group, Michigan State University professor Frank Child told Gordon that it was "in the political and administrative area" that a solution to the nation's instability might be found. Baldly separating economic and political considerations like these came as second nature to this professional social scientist. Similarly, Gordon and his Council colleagues believed that national economic policy could be formulated on a rigorous basis in ways cognizant of, yet not contaminated by, ideology and political faith. For them all, and for the profession they represented, it was a settled judgment that would in the end yield only disappointment. Quite understandably and in his own way, Child anticipated the harsh events that would make a mockery of that shared conviction;

before writing Gordon he had sent his wife and children to Europe having become "a little nervous about the bombings and terrorist activities" in Saigon.[33]

Notes

1. See, for example, Michael J. Webber and David L. Rigby, *The Golden Age Illusion: Rethinking Postwar Capitalism* (New York: Guilford Press, 1996); Stephen A. Marglin and Juliet B. Schor, eds., *The Golden Age of Capitalism* (Oxford: Clarendon Press, 1990); Michael A. Bernstein, "Understanding American Economic Decline: The Contours of the Late-Twentieth-Century Experience," in *Understanding American Economic Decline*, ed. Michael A. Bernstein and David Adler (New York: Cambridge University Press, 1994), 3–33. See also Harold Vatter and John Walker, eds., *History of the U.S. Economy since World War II* (Armonk, N.Y.: Sharpe, 1996). Needless to say, military spending itself provided another source of growth for the economy as a whole. See, for example, George Hildebrand et al., "Impacts of National Security Expenditures Upon the Stability of the American Economy," in *Federal Expenditure Policy for Economic Growth and Stability* (Joint Economic Committee of the Congress [November 5, 1957]; Washington, D.C.: Government Printing Office, 1957), 523–541.
2. See Paul A. Samuelson, *Economics* (New York: McGraw-Hill, 1972), 250. First published in 1948, the Samuelson text remains in print to this day, thirteen editions and millions of sold volumes later. President Johnson's declaration is found in Council of Economic Advisers, *Economic Report of the President: 1965* (Washington, D.C.: Government Printing Office, 1965), 10.
3. The three slumps notwithstanding, weekly earnings for American workers rose throughout the decade. Living standards for most greatly improved. A clear majority of the nation's population worked shorter hours and enjoyed the amenities of a much broader array of durable and nondurable goods than their forebears. Before World War II, for example, approximately a quarter of the nation's farm households had electricity; by 1959, over 80 percent of them did, and, as a consequence, they also had telephones, televisions, and refrigerators. See United States Department of Labor, *Economic Forces in the U.S.A.* (Washington, D.C.: Government Printing Office, 1960), 73, 80; U.S. Department of Commerce, *Statistical Abstract of the United States* (Washington, D.C.: Government Printing Office, 1960), 336.
4. On all these matters, see, for example, John Morton Blum, *Years of Discord: American Politics and Society, 1961–1974* (New York: Norton, 1991), 20–21, who also notes that "Kennedy's narrow victory, as Nixon saw it, owed more to dissatisfaction with the economy than to the Democratic campaign"; Stephen E. Ambrose, *Eisenhower: The President* (New York: Simon and Schuster, 1984), as well as his *Nixon: The Education of a Politician* (New York: Simon and Schuster, 1987); Arthur Schlesinger Jr., *A Thousand Days: John F. Kennedy in the White House* (Boston: Houghton, Mifflin, 1965); John Patrick Diggins, *The Proud Decades: America in War and in Peace, 1941–1960* (New York: Norton, 1988), 340–343; and Theodore H. White, *The Making of the President: 1960* (New York: Atheneum, 1961).
5. The original members of the Kennedy CEA were: Kermit Gordon of Williams College, a Swarthmore alumnus and former Rhodes Scholar; Walter Heller (the Council chair) of the University of Minnesota, who had received his professional training at the University of Wisconsin, been chief of internal finance for the U.S. Military Government in Germany, and served on the Economic Cooperation Admin-

istration mission to Germany; and James Tobin of Yale University, who had trained at Harvard University, received the distinguished John Bates Clark Medal of the American Economic Association in 1955, and served as president of the Econometric Society in 1958.

6. See James Tobin, "Academic Economics in Washington," *Ventures* 2 (winter 1963): 24–27 at 26.
7. The words are John Morton Blum's from his *Years of Discord*, 56.
8. That the activist posture of the Kennedy Council bore striking similarities to that of Truman's is candidly discussed in an exchange of letters between Leon Keyserling and James Tobin in the early 1970s. See Tobin to Keyserling, August 5, 1971, and Keyserling to Tobin, July 21, 1971; both in the Papers of Leon Keyserling (PLK), Harry S. Truman Library, Independence, Missouri, Box 10, Folder: "Tobin, James—Keynesian Economic Theory." The issue is further discussed, along with observations concerning Truman's lack of engagement with the work of the Council, by Keyserling. See "Oral History Interview with Leon H. Keyserling," by Jerry N. Hess, Washington, D.C., May 3, 10, 19, 1971, Harry S. Truman Library, 161–172, 190–194. On Kennedy's "analytical mind," see "Oral History Interview with Walter S. Salant," by Jerry N. Hess, Washington, D.C., March 30, 1970, Harry S. Truman Library, 44–45.
9. During the 1960s and 1970s, the American share of real national product devoted to defense procurement and operations was almost double that of West Germany and (not surprisingly, given the provisions of the Security Treaty) close to six and a half times larger than that of Japan. I discuss the mechanisms accounting for the economic renaissance of those nations devastated by World War II, and their consequence for American economic performance, in "Understanding American Economic Decline," 16–24. By the late 1970s, it was not uncommon for American workers, especially in the automobile, steel, and textile industries, to employ equipment several decades older than that engaged in foreign production. The consequences in poor productivity growth were painfully obvious. See Michael A. Bernstein, "Economic Pessimism and Material Prosperity," in *The Humanities and the Art of Public Discussion*, vol. 2 (Washington, D.C.: Federation of State Humanities Councils), 19–27.
10. A succinct and straightforward narrative of these early debates over economic policy in the Kennedy cabinet is offered by Blum in *Years of Discord*, 53–57.
11. Needless to say, the public deficits envisioned by the Kennedy CEA as a necessary policy instrument had the potential to encourage the Fed in its pursuit of a "tight" monetary stance. Larger federal government borrowing requirements would, in the view of most members of the Board of Governors, require higher interest rates as the relative supply of loanable funds to the private sector was reduced in order to meet public sector needs. Such "disintermediation" in the nation's capital markets was (and is) another oft-repeated explanation for the tendency of fiscal and monetary policy to clash.
12. A particularly clear and concise statement of the fiscal objectives of the Kennedy Council is provided by James Tobin's "Growth through Taxation" (originally published in the *New Republic*, July 25, 1960) in *National Economic Policy* (New Haven, Conn.: Yale University Press, 1966), 78–88.
13. See "Oral History Interview with Walter Heller, Kermit Gordon, James Tobin, Gardner Ackley, and Paul Samuelson," by Joseph Pechman, Fort Ritchie, Md., August 1, 1964, John Fitzgerald Kennedy Library, Boston, 34, 36–37, 72–73.
14. See ibid., 62.
15. See Arthur M. Okun, "Potential GNP: Its Measurement and Significance," in *The*

Political Economy of Prosperity (Washington, D.C.: Brookings Institution, 1966), 132–145.

16. See Heller to Nourse, March 4, 1966, Records of the Council of Economic Advisers: 1961–1968 (ROCEA: 1961–1968), Lyndon Baines Johnson Library, Austin, Texas, Microfilm Roll 36, Folder: "Walter Wolfgang Heller (WWH)—thru October 1965"; as well as Keyserling to Lyndon Baines Johnson, May 25, 1963; Heller to Keyserling, April 30, 1963; Keyserling to Heller, May 13, 1963—all in ROCEA: 1961–1968, Microfilm Roll 37, Folder: "Leon Keyserling." See also Jacoby to Burns, July 17, 1967, and Burns to Jacoby, August 3, 1967—both in Arthur F. Burns Papers (AFBP), Dwight D. Eisenhower Library, Abilene, Kansas, Box 12, Folder: "Neil J. Jacoby (1)"; Mordecai Ezekiel (then of the Food and Agriculture Organization) to Arthur Smithies (Harvard University), January 13, 1962, Papers of Mordecai Ezekiel (PME), Franklin D. Roosevelt Library, Hyde Park, New York, Box 7, "Ezekiel, Mordecai: Personal, 1962–63."
17. In mid-August 1961, the German Democratic Republic—under orders from Moscow—erected a wall dividing the eastern and western zones of occupation of Berlin. This direct contravention of the agreement among the allied powers, in the wake of World War II, to preserve open access to the entire city brought the military forces of both the United States and the Soviet Union to full alert status. While the immediate crisis passed without incident, save for the killing of many East Berliners who sought to cross the line of demarcation to the West, it led to the expansion of the American garrison in central Europe for decades to come. On the $3–4 billion projection and Kennedy's anxieties about it, see "Oral History Interview with Walter Heller, Kermit Gordon, James Tobin, Gardner Ackley, and Paul Samuelson," 377–378.
18. See ibid., 179, 219–220. See also Heller to Robert F. Wallace (School of Business Administration, Montana State University), June 7, 1963, PWWH, Box 50, Folder: "W-Wehrle"; "Oral History Interview with Arthur M. Okun," by David McComb, Washington, D.C., March 20, 1969, Lyndon Baines Johnson Library, 20, 5–11, 15.
19. Arthur Okun ultimately credited Heller with winning the president's support in the struggle over possible tax increases associated with the Berlin standoff. See "Oral History Interview with Arthur M. Okun," 12.
20. On the creation of the Federal Reserve System, the most recent and thorough contribution is James Livingston, *Origins of the Federal Reserve System: Money, Class, and Corporate Capitalism, 1890–1913* (Ithaca, N.Y.: Cornell University Press, 1986). See also Michael A. Bernstein, "The Contemporary American Banking Crisis in Historical Perspective," *Journal of American History* 80 (1994): 1382–1396.
21. The determination of the Fed Board of Governors to maintain their independence in the later postwar period stemmed in large measure from debates that emerged after 1945 concerning interest rate policy. During both the Great Depression and World War II, the Fed had been obliged to support Treasury bill issues at par, the objective of the arrangement being to keep rates of return low enough that the market value of Treasury securities would remain high and the servicing of the national debt made accordingly easier. Yet this task tied the hands of the monetary authorities when and if they sought to raise interest rates in the hopes of stemming inflation. The Fed governors became particularly aggrieved as the price level rose after wartime market controls were eliminated; in their view, the Treasury bill policy left them powerless to control an upward spiral in the cost of living. Just before Harry Truman left office, in what became known as the "Accord of 1951," the Fed was released from its obligations to the Treasury—a step that

both freed monetary policy to pursue a more aggressive anti-inflationary course and fostered among the governors an enduring spirit of independence and authority in broader issues of economic policy as a whole. James Tobin comments on these developments in *New Economics*, 12–13, as well as in "The Future of the Fed," (originally published in *Challenge* [January 1961]), in *National Economic Policy*, 134–143.

22. See Tobin, "The Future of the Fed," 136. On Martin's surprising refusal to submit his resignation, see "Oral History Interview with Walter Heller, Kermit Gordon, James Tobin, Gardner Ackley, and Paul Samuelson," 188–189, in which Tobin speculated that a large part of the Fed chair's reluctance had to do with his suspicion that his pro forma submission would in fact be accepted.
23. See "Oral History Interview with Walter Heller, Kermit Gordon, James Tobin, Gardner Ackley, and Paul Samuelson," 136, 195. On the delicate diplomacy that Walter Heller thought essential "[to] the interest of getting the maximum cooperation from the 'independent' Federal Reserve," see Heller to Paul H. Douglas (Democratic senator from Illinois), July 10, 1961, PWWH, Box 47, Folder: "Copies of Outgoing Letters 3/24/61–7/24/61." In forging the modus operandi that eventually emerged between the White House and the Fed, John Kenneth Galbraith evidently played an important role—especially in a personal appeal to William McChesney Martin. See Heller to Galbraith (then ambassador to India), August 6, 1961, PWWH, Box 47, Folder: "Copies of Outgoing Letters 8/5/61–11/14/63."
24. See "Oral History Interview with Walter Heller, Kermit Gordon, James Tobin, Gardner Ackley, and Paul Samuelson," 326, 328, 330–331. Heller recalled that the Quadriad was originally known as the Financial Summit Meeting. Having coined the term "Troika," Heller did not know the Russian equivalent for foursome—*chetvyorka*. See also "Oral History Interview with Charles L. Schultze," by David McComb, Washington, D.C., March 28, 1969, Lyndon Baines Johnson Library, I–7. Formation of the Troika and the Quadriad is also discussed in the *Administrative History of the Council of Economic Advisers* (AHCEA), Box 1, Folder: "Chapter 1: The Council as an Organization," 14–15, and appendix including a copy of a memorandum from Heller to President Johnson, December 1, 1963, 3–4, Lyndon Baines Johnson Library.
25. See Max M. Kampelman to Heller, April 5, 1961, and Heller to Kampelman, April 24, 1961, PWWH, Box 49, Folder: "K"; and Walt W. Rostow (special presidential assistant for national security affairs) to Henry R. Labouisse (director, International Cooperation Administration), Box 3, Folder: "WA–WH," March 13, 1961, Papers of John F. Kennedy, Presidential Papers–White House Staff Files, Walt W. Rostow File, John Fitzgerald Kennedy Library. Samuelson reports on Johnson's idea and on his own reaction to it in "Oral History Interview with Walter Heller, Kermit Gordon, James Tobin, Gardner Ackley, and Paul Samuelson," 366–367.
26. See "Oral History Interview with Gardner Ackley," by Joe B. Grantz, Ann Arbor, Mich., April 13, 1973, Lyndon Baines Johnson Library, II–20–1.
27. Heller was later quite forthcoming about his ulterior motives in the Quadriad's formation in "Oral History Interview with Walter Heller," by David McComb, Minneapolis–St. Paul, February 20, 1970, Lyndon Baines Johnson Library, II–1–2. While the Bureau of the Budget did not have formal cabinet status during the Kennedy administration, it had understandable influence over White House decision-making. Reorganized as the Office of Management and Budget in 1971, its director was formally given cabinet status by President Richard Nixon. On the effective operations of the Quadriad, see "Oral History Interview with Walter

Heller, Kermit Gordon, James Tobin, Gardner Ackley, and Paul Samuelson," 329.

28. On Tobin's chat with Sorensen, see "Oral History Interview with Walter Heller, Kermit Gordon, James Tobin, Gardner Ackley, and Paul Samuelson," 235. Okun's observation is made in "Oral History Interview with Arthur M. Okun," 14.
29. See Roy E. Moor (administrative assistant to William Proxmire) to Heller, October 10, 1964, PWWH, Box 49, Folder: "M"; Burns to Heller, May 11, 1964, AFBP, Box 12, Folder: "Council of Economics Advisers—Walter W. Heller"; Friedman to Heller, December 30, 1960, and Rashi Fein (Brookings Institution) to Heller, September 16, 1963, both in PWWH, Box 48, Folder: "F"; Marcus Alexis (DePaul University) to Heller, January 19, 1961, PWWH, Box 47, Folder: "A"; and Leroy S. Wehrle (U.S. AID Mission to Laos) to Heller, July 8, 1964, PWWH, Box 50, Folder: "W–Wehrle."
30. Heller's central role in the emergence of the antipoverty strategy is attested to by an array of testimony from members of the administration. Then director of the Bureau of the Budget (and future CEA chair under President Jimmy Carter), Charles Schultze, who had noted the president's high regard for Harrington, also credited Heller as the individual "who probably first put the idea [of a poverty policy] in Kennedy's head." See "Oral History Interview with Charles L. Schultze," by David McComb, Washington, D.C., March 28, 1969, Lyndon Baines Johnson Library, I–38–9. Similar recollections are provided by both Gardner Ackley and Robert Lampman. See "Oral History Interview with Gardner Ackley," I–9; "Oral History Interview with Robert Lampman," by Michael L. Gillette, Madison, Wis., May 24, 1983, Lyndon Baines Johnson Library, I–1–3. As for Harrington's work, the president had evidently read his most recent book, *The Other America: Poverty in the United States* (New York: Macmillan, 1962). A CEA staff member, Robert Lampman had generated the estimates of skewed wealth and income distribution (exhibited in data drawn from the early 1960s) that so struck Heller. Lampman remembered ("Oral History Interview with Robert Lampman," I–4) that Heller specifically asked that a "simplified version" of the econometrics memorandum be prepared for the president's consideration. A graduate school classmate of Heller's and later a professor of economics at the University of Wisconsin–Madison, Lampman wrote a chapter of the *Economic Report of the President: 1964* dealing specifically with the proposed War on Poverty. When Lampman died in 1997, James Tobin called him "the intellectual architect of the war on poverty." See *New York Times*, March 8, 1997.
31. From Walter Heller's "Confidential Notes on Meeting with the President, October 21, 1963," 2, PWWH, Box 13, Folder: "10/18/63–10/31/63." Heller indicated that his briefing of the president followed up on a "memo of last summer." Regarding the links between the tax cut policy and efforts to end employment discrimination, see Lloyd Ulman (director, Institute of Industrial Relations, University of California–Berkeley) to Heller, July 3, 1963, PWWH, Box 12, Folder: "7/2/63–7/20/63." On Heller's meeting with the president before the trip to Dallas, see "Oral History Interview with Robert Lampman," I–14.
32. See Heller's "Notes on Meeting with President Johnson, 7:40 p.m., Saturday, November 23, 1963," 1–3, PWWH, Box 13, Folder: "11/16/63–11/30/63." Regarding the flight from Dallas to Washington on *Air Force One* and Heller's preparation for the meeting with Johnson, see "Oral History Interview with Joseph A. Pechman," by David G. McComb, Washington, D.C., March 19, 1969, Lyndon Baines Johnson Library, 24.
33. See Child to Gordon, March 1, 1961, Papers of Kermit Gordon, John Fitzgerald Kennedy Library, Box 28, Folder: "Correspondence (KG)." In the wake of the 1954

Geneva Accords that divided the Democratic Republic of Vietnam (to the north) from the Republic of Vietnam (to the south), the United States formed the Military Assistance Advisory Group (MAAG) to aid the South Vietnamese government of Ngo Dinh Diem. By the middle of 1961, as the American army garrison expanded to 23,000 soldiers, the MAAG became the Military Assistance Command–Vietnam.

Part II

The Social Sciences as Process and Procedure

WHETHER THERE HAS BEEN any enduring content to the social sciences or whether they are all process and procedure is a major question insofar as science in general is concerned and, for that matter, any other artifact of human culture. Although that is a question about which honest people will continue to disagree longer after this book's contribution to the history of the social sciences is exhausted, it is of great importance to weigh in on the matter. As late as World War I, or even beyond that date, many educated Americans—especially businessmen and scientists in the well-established physical and biological sciences—expressed doubts in public and in private that the social sciences were competent fields of knowledge, without distinctive or rigorous methods, an agreed-upon content for each discipline, and major controlling hypotheses or theories; in short, they were not sciences or disciplines worthy of the name. As late as the early 1960s physicist-cum-philosopher-of-science Thomas S. Kuhn insisted that the social sciences were just out of the "preparadigmmatic" phase of their development, thus suggesting, with a left-handed compliment, that among the sciences, they were very much the newest kids on the block, with much to learn from their older and more accomplished competitors in such fields as physics, chemistry, and biology.

Hence it was to be expected, perhaps, that as the social sciences organized themselves in research and land grant universities in the interwar years, that their leaders and champions within and without the social disciplines would emphasize how scientific, how rigorous, how powerfully did their work illuminate social issues and problems. In particular did two major contributions to method make a difference. The first was quantification, and certain leaders, such as Franz Boas in anthropology, Edward Lee Thorndike in psychology, William F. Ogborn in sociology, Wesley Clair Mitchell in economics, and the students of Charles E. Merriam in political science, led the way in the 1910s and 1920s. By the 1930s quantification and statistical methods came to the social sciences and helped nourish a virtual religion of scientific positivism among many social scientists coming of age in the profession in the 1930s and 1940s. The second methodological contribution was behaviorism, or the behavioral revolution, which the psychologists really began in the 1910s by shifting their foci from the traits of the organisms they studied to their patterns of behavior as members of particular groups; indeed, psychology's founder of behaviorism, John Broadus Watson, was interested precisely in the patterns of behavior of, say, an *average* monkey or cat or human infant. For him and most other American social scientists, individuals did not exist outside of groups; the only social reality was the group, not the individual. As behaviorism spread to sociology, anthropology, economics, and other social sciences, it became clear that behaviorism was linked to various forms of logical positivism and determinism. It was no small irony that during the 1930s, 1940s, and 1950s, as ordinary Americans responded to the threats of European and Asian totalitarianism to preserve the freedom of the individual, that American social science, often enlisted in these moral struggles, had undermined the scientific legitimacy of such political beliefs. In short, there was a latent, and sometimes a not so latent, tension between the nation's political and scientific ideas.

The authors of the chapters in Part II all follow up the tensions, in one way or another, between American politics and American social science, especially as science policy. Did the social sciences transmit social values and political priorities, as their critics have often complained? Or have they functioned as mere tools in the hands of whatever politico or apparatchik might wrap his or her grimy mitts around them for his or her own purposes? Does method—or means—get the better of ends—or goals and values? What sort of outcomes can one expect from such technical perspectives and points of view?

In a slender but nevertheless suggestive piece, Harvey M. Sapolsky argues that we have evolved a new method of argumentation and presentation within our military that has made it an all-powerful institution whose architects have virtually restricted the ways in which the federal government functions and thus how military and science policy are made. World War II was a war of attrition, in which the major belligerents used a Mass Army Heavy military format of battle and operations. The real Wizard War was not that conflict, Winston Churchill's views notwithstanding, but the Cold War. Beginning with the anti-U-boat campaign in the North Atlantic, in which American forces were initially inferior to Axis power, Allied military began to emphasize the scientific analysis of warfare components, or the military format of

Thorough Exploitation of Science and Technology (TEST), meaning that in this format the military looks for doctrinal, equipment, and training improvements that will yield an edge in combat always sufficient to stop any enemy's effort to overwhelm through sheer numbers of soldiers, equipment, and resources. An early example of this new strategy was the atomic bomb. As this strategy spread in the military in the early years of the Cold War to other operational units of our armed forces, TEST became what distinguished our strategy and tactics from that of our rivals, the nations of the Soviet bloc.

But more was involved. The military used quite a bit of social science in World War II and in the Cold War, but no innovation proved more important than the Planning, Programming, and Budgeting System (PPBS), which President Kennedy's defense secretary, Robert S. McNamara, borrowed from the RAND corporation in the early 1960s. In McNamara's hands, the new systems analysis became a tool for decision-making, especially in the acquisition of weapon systems, and an instrument that befuddled the traditional officer cadre until they could have their own experts in the technique. By then, McNamara and his successors had won many battles. How systems analysis became an important tool in military and domestic decision-making—a methodological contribution from social science if there ever was one—is a story that cannot be told in all its complexities in these pages; but this is a good point of departure for further investigation.

In "A Risk Perceived Is a Risk Indeed: Assessing Risk in Biomedical Research and Health Policy," Philip L. Frana insists that there is an underlying conflict between that most disturbing phenomenon of American culture since the 1950s, that of cultural fragmentation and individuation and consequent democratic and individualistic decision-making, which have changed the very way in which we think and act about health risks. Bluntly put, once upon a time we believed in public health, in a shared body of concerns and issues and methods of dealing with them, based on an actual and real social community. Now there is no such thing as public health, but only the personal health of private individuals in their own private lives. Health assessment is in our own time a mushrooming area of concern for the federal government. In that sense, few social sciences besides public health have generated as much interest and political weight; in 1993 the now-defunct Office of Technology Assessment estimated that health assessment research totaled half a billion dollars in the federal budget, and as a nation we probably spend trillions of dollars a year in pursuit of better health. Cost containment has come to mean personal responsibility for a more healthy lifestyle that each and every individual was supposed to pursue. No longer were operational the older communal models of equal risk and equal responsibility for all individuals in a particular category. Once the notion of community evaporated, so did the conception of a shared public health, with the ironic result that no two individuals could necessarily have equal outcomes. All individuals are different. All get different results. Risk, then, Frana concludes, is a cultural and intellectual artifact, and what we see in it and what we think we can get from it are what matter the most. In other words, social science (in this case public health) does not trump the fundamental assumptions and procedures of American society and culture but derives its intellectual and social configurations from that larger society.

In the final chapter in this section of the book, Howard P. Segal carries Frana's analysis even further. In an adept sketch of the rise of social science policy in the federal government, Segal shows how certain social scientists and intellectuals in the 1950s and 1960s made such a possibility respectable and even a positive good, to counter the skepticism and contempt that physical scientists such as Vannevar Bush and others displayed toward the social sciences in the immediate post–World War II years. In particular did the social scientists make the argument that the use of social science would make society better—more prosperous, healthier, more productive, and less afflicted with a host of serious social problems. Above all, certain social science ideas and notions, especially so-called modernization theory, became the model for politicians and social scientists alike in the Cold War seeking to make the Third World safe for American foreign policy and exports. The transfer of American machinery, seeds, and social engineering to such places would turn their governments away from the temptations of Soviet Communism and toward Western capitalism. And American policy-makers and social scientists came to believe, especially in the 1960s and 1970s, that public policy at home and in the world were not too much for the right mixture of experts to help manipulate and commandeer for their political masters. As Segal brilliantly argues, the architects of this Cold War–era systems engineering "improved" on earlier versions of the method by ignoring the system's dynamics and, indeed, the system itself and looked only at the final or end product, enabling such analysts to focus on any element from a single perspective: what was its output, and how could that output be increased? Each part was now an individual part rather than, as in early systems theory, a part of the whole that did not separate from the whole. Thus different kinds of experts were needed, and social scientists were as necessary as anyone else. But the whole had been individuated, which was the road to disaster, for that perspective allowed no unified vision of policy or any part of it. There was instead a multitude of equally expert voices—chaos indeed. And all had come full circle from Bush's *Science—The Endless Frontier* (1945) to the Clinton administration's, *Science in the National Interest* (1994), whose authors plumped for expanded federal support for science and technology in the post–Cold War era. After some specious shadowboxing about the relations of applied and basic research, *Science in the National Interest* celebrated recent advances in high technology while refusing to recognize a number of failures, for that would have meant recognizing that for a growing number of American citizens, science and technology—including social science—have lost much of their moral and intellectual authority, not to mention the deference and reverence formerly awarded them by many citizens. For many Americans, the advantages of science and technology have devolved to the point where they are chiefly prized as expressions of individual expression, as in downloading music onto one's computers, buying stocks on the Internet, or even finding old girlfriends or boyfriends through the marvels of *Google*. Once again, American culture and society's fundamental trends and assumptions seem to trump science and technology, at least to the extent of causing a crisis in their public legitimation.

Chapter 3

The Science and Politics of Defense Analysis

HARVEY M. SAPOLSKY

THE MOBILIZATION for war was so frequent, so large, and so enduring during the twentieth century that it is difficult to find a sector of American society that has been untouched by it. The Cold War alone spanned forty-five years and cost approximately $14 trillion. During this period the United States never had less than two million men and women under arms, and sometimes more than three. One million more served simultaneously in the reserves. Another three or four million worked as civil servants or contractors on defense-related projects. Defense in the twentieth century came to involve the natural sciences, engineering, the social sciences, medicine, education, communications, and nearly everything else associated with modern life.[1]

Although it is appropriate to be concerned about any distortions the Cold War mobilization may have imposed upon segments of American society, we have to recognize how light the war's touch was in fact. Despite the billions invested in weapons, it was the Soviet economy, not the American economy, that ultimately collapsed.[2] Despite the number of people in uniform and the exploitable fears of foreign subversion, Americans did not lose their freedoms. And despite the need to centralize elements of political decision-making to deal with the threat of sudden nuclear attack, our system of checks and balances on political power survived intact.

The way to understand the role the social sciences played in the Cold War is to contrast the experience of the natural sciences and engineering with that of the social sciences. Not all disciplines are treated equally within the universities or gain the same role in the nation's defense. The Cold War military was a very wealthy and generous patron of research, but also sometimes a very opinionated and demanding one.

Winston Churchill called the Second World War the Wizard War to underline the importance of science in the Allies' success. But in reality it was not science but industrial production and engineering that won the war. America was the "Arsenal of Democracy," producing 300,000 aircraft, 100,000 armor vehicles, and 20,000 ships, many more than did its enemies and enough to supply its military and those of its allies. The Second World War was a war of attrition, where metal clashed against metal. Allied weapons were not always better than those of the Axis powers, but we had many more than they did.

Science and Technology in the Cold War

The Cold War was different. Advances in science and technology were the keys to victory and recognized early as such. The Second World War is described in military format terms as a Mass Army Heavy, a war of mass-produced tanks and mass-produced soldiers.[3] The Soviet Union kept to that format during the Cold War; the United States did not. Beginning with the antisubmarine campaign in the Atlantic during the Second World War and spreading gradually to other segments of the military—strategic forces in the 1950s, tactical air in the 1960s, and then armor forces in the 1970s—the U.S. military came to emphasize the scientific analysis of warfare components, what Barry Posen calls the Thorough Exploitation of Science and Technology (TEST) military format.[4] In this format the search is for doctrinal, equipment, and training improvements that will yield an edge in combat sufficient always to negate any attempt by an enemy to overwhelm through numbers.

The military initially resisted offering scientists and engineers an increased role in the development of weapons and guiding the use of military force. At the beginning of the Second World War, senior officers feared that scientists, and especially the scientists advocating early American entry into the war who led the preparedness movement, were too closely tied to the British and were likely to demand too much independence. Worse, they thought that scientists had little to contribute to the war effort itself. If scientists wanted to help, they were told to put on a uniform and implement development plans the military had already outlined. But by the end of the war the value scientists provide in developing new weapons and designing improved tactics was obvious to the military. Some sort of accommodation with scientists was necessary if the nation was to maintain its military strength in the future.[5]

The problem was that the scientists most interested in continuing work on defense-related problems indeed did prefer independence from the military. The hierarchy and discipline characteristic of military-managed organizations had little appeal to those returning to their laboratories after the war.

More important, Vannevar Bush and the others who led the wartime mobilization of science remembered well the conflicts they had with senior officers whom they considered insufficiently appreciative of the contributions that scientists had to offer. Led again by Bush, the scientists made three unsuccessful attempts to secure an autonomous base. They proposed in sequence establishing a Research Board for National Security at the National Academy of Sciences with a billion-dollar federal endowment to conduct weapons research independently of the military, a national security division in what became the National Science Foundation, and giving the Joint Research and Development Board, which Bush headed, status within the Department of Defense equivalent to that of one of the armed services. The latter initiative was especially threatening to the military because it would have given Bush and his successors a seat on the Joint Chiefs of Staff, an exclusive military preserve.[6]

The military had no interest in sharing power directly with scientists but did need their involvement in the development of modern weapons. The services wanted knowledgeable, trusted technical advisers—people who understood military problems but also the fact that policy decisions are shaped by many factors, technology being only one of them. Several obvious avenues for assistance were deemed inadequate. Civil service positions, reflecting a deep cultural bias in the United States, could not attract the nation's top technical talent. Industry clearly could offer scientists and engineers significant financial rewards but in the end requires a focus on profits rather than on the public interest. Universities provide the independence that scientist value so much, but it was this independence that worried the military. The solution was to create a set of not-for-profit organizations dedicated to work on defense problems and often managed by universities but dependent upon the military for their financial support.[7] The RAND corporation, Lincoln Laboratory, Los Alamos National Laboratory, the Aerospace corporation, and the Institute for Defense Analyses are among the dozens of such institutions. They pay better than the civil service, offer a university-like work environment, have privileged access to government data, but do not challenge openly government policy and plans because the government is their sole source of support. They have what Sanford Weiner once called constrained autonomy.[8]

The military's interest in supporting the weapons research of scientists and using scientists as advisers was to best the Soviet Union. In a very real sense scientific research itself was also a weapon. The continuous ability of America to develop superior weapon systems was viewed as a major threat by the Soviets and recognized as such by both the American and the Soviet militaries. A substantial part of the justification within the government for projects like the Strategic Defense Initiative (Stars Wars, to its detractors) was that they would strain the Soviet Union psychologically and potentially bankrupt it if a match were attempted, as was likely.[9] Each new exotic project would underline the technological gap that existed between the adversaries, creating self-doubt among the Soviets about their ability to compete.[10] It hardly mattered that most were beyond our own ability to realize quickly or at a reasonable cost.

The Social Science Weapon System

Social science was also viewed as a weapon, but for use against a different enemy. For the military, the most useful social science available during the Cold War was that embodied in policy analysis, and policy analysis was used as argumentation in domestic bureaucratic battles for budget and missions. The enemies then for the navy were the air force, the army, the Office of the Secretary of Defense, Congress, the president, and all the others who needed to be influenced if the navy were to get its way in policy wars. As much for defense as for offense, each of the armed services developed a set of social science–dominated research organizations to argue its positions. In this sense, RAND, which initially worked exclusively for the air force, begat the Center for Naval Analyses, which begat ANSER, a think tank that worked for the army, and the Institute for Defense Analyses, which advised the Joint Chiefs of Staff.[11] It is important also to note that much of this work has been absorbed within the military itself over time rather than being further contracted out, as has the military's natural science and engineering expertise. Unlike technical advice, policy advice is too much aimed at the organization's core interests, budgets, and jurisdiction, to be farmed out.

To be sure, as a giant among large-scale bureaucracies, the military has long supported research in certain social science subfields—occupational testing, job satisfaction surveys, troop morale, and training techniques, for example—that might be useful in improving managerial effectiveness.[12] Facing the potential of very rapid expansion as in the case of war, military leaders want to know how best to prepare and place hundreds of thousands of recruits. There is also a need to understand the motivation of soldiers to fight. Is it for the national cause or for foxhole buddies? And the military needs to devise systems for training that can quickly prepare recruits for demanding positions. All of this is relevant to leaders in other large organizations who need assurance that their newly hired pegs will fit standard holes without disrupting production or customers. In this sense the army, General Motors, and Wal-Mart have very similar needs.[13] There is also a strong military interest in area studies, work on particular regions where military operations are likely to occur.

Much like an international bank or development agency, the military believes it does better when it knows the language, culture, and terrain when the time comes to venture forth. MIT's Center for International Studies, my own home institution, is a product of this interest, though by the State Department and the Central Intelligence Agency rather than by one of the armed services.[14]

The best example of the bureaucratic leverage social science provided during the Cold War is that of systems analysis employed by Robert S. McNamara, secretary of defense for seven years during the Kennedy and Johnson administrations.[15] Despite the personal and national anguish produced by presiding over a lost war, Secretary McNamara is still viewed as one of the greatest administrators ever to have served in American government. It is a reputation largely built on the claimed effects of system analysis techniques and the complementary Planning, Programming, and Budgeting System (PPBS), both of which were brought into the Defense Department by McNamara from RAND, where they had been developed by economists and political scientists. A definite sign of their political success, though not necessarily of their administrative efficaciousness, was the rapid diffusion of systems analysis and PPBS to other government departments.[16] Years after both their introduction and the exposure of their substantial limitations, they remain hallmarks of good government.

The introduction and limitations of systems analysis and PPBS are described crisply in Arnold Kanter's wonderful *Defense Politics*.[17] The Kennedy administration, Kanter notes, had gained office in part on claims that the Eisenhower administration had mismanaged the nation's defense. The argument was that Eisenhower had placed an arbitrary ceiling on defense expenditures rather than examining weapon development and force structure proposals exclusively on their potential contribution to defense. This approach, the Democrats claimed, necessarily led to an emphasis on nuclear weapons, which were powerful and relatively cheap but also very awkward indeed to use. Kennedy favored the so-called Flexible Response strategy, which was built around the graded escalation of violence and thus conventional weapons. McNamara embraced the new strategy and promised somewhat disingenuously a cap-free budget process for the services.

Despite the fact that he was the five-star general who led the crusade to free Europe in the Second World War, Eisenhower did not control defense spending during the intense early years of the Cold War by pulling rank or combat experience on his former comrades in arms. Rather, he relied on his popular election as president to select a budget ceiling for defense that would not overly burden the nation's economy. If the service chiefs wished to debate the cap he had chosen, Eisenhower was in essence saying go out and win election yourself. McNamara was the agent for Kennedy and Johnson, who did in fact win election, but how then could he find a way to limit defense spending, as he must if the new administration was to have a domestic agenda? He could not establish a visible cap on defense spending because that was what the Democrats had accused the previous administration of doing. Nor could he easily contest directly the programs the chiefs might propose on the grounds that his military knowledge or experience exceeded theirs. It was then that he reached to RAND for a methodology for analyzing alternatives and making budget decisions, systems analysis and PPBS, for which civilians were the experts and not the military.

Systems analysis is little more than systematic thinking mixed with some attempt at problem quantification.[18] Try, its proponents tell us, to consider and rank in priority all the parameters in evaluating program alternatives. Seek out real or surrogate measures for as many of the parameters as possible. PPBS is a budgeting system that seeks to group the budgets of likes with likes even though they may be individually the responsibility of different organizational subunits. An example would be to group all of our strategic forces—bombers, land-based ballistic missiles, and submarine-based ballistic missiles—the organizational responsibility of several different commands and two services. Coupled with a schedule of formal decision points, it becomes a framework for managing budgets and programs together rather than separately. Both techniques had a following among some academics and administrative theorists, with PPBS's origins dating back to the 1920s and systems analysis applied to military problems to the 1940s, but they were essentially unknown to senior practitioners until their importation from RAND seminar rooms to the halls of the Pentagon in 1961.[19]

In order to control the defense budget, the Office of the Secretary of Defense had to choose among a number of acquisition and deployment programs, each one of which had its own military advocates and military rationale. Faced with a medal-bedecked general or admiral arguing that the fate of Western freedom depended upon this program and this program alone, it was hard for McNamara and his staff to prevail. But once the discussion was made to focus on, say, the combat values of different sortie rates generated by different types of aircraft and other measures of the analysts' making, power slipped into the hands of the civilians. At least initially, military officers were completely unprepared to debate in the new language or otherwise cope with the cadre of self-confident system analysts the secretary assembled to enforce his writ within the department. Officers noted unhappily that some programs were subject to intense analysis while others, more sensitive politically or especially favored by the administration, were not. They grumbled about the arrogance and inexperience of the young analysts (McNamara's Whiz Kids) and lost nearly all the battles they had with them.[20]

But the military learns quickly to adapt to new terrain and tactics. Soon young officers were off to universities offering courses in systems analysis and program budgeting techniques to earn degrees in public policy. Staffs were assembled to coordinate studies conducted by the service's dedicated not-for-profit research organizations. These organizations countered the studies that the secretary's staff prepared and those generated by the captive think tanks of the other services. Promotions and choice assignments were available for officers who mastered the new art but who stayed loyal to their service.

What was special and effective has become commonplace and largely reflexive. Now nearly everyone knows enough about the policy game to search for the system parameter that uniquely favors his or her preferred outcome. Many can create the numbers and do the calculations that prove that theirs is the cost-effective alternative. The advantage of touting system analytic techniques fades, but not the search for an advantage. Today it lies in model building and computer simulations. Tomorrow it may lie in content analysis or psychological profiling.

Doing Science In-house or Out

Governments have the choice of creating internal capacity to conduct research and development activities or contracting with profit-making or not-for-profit private organizations for the work. As was described previously, the American military preferred to utilize nongovernmental organizations for much of its weapon development effort during the Cold War. Government facilities (arsenals, laboratories, test ranges) were thought to be both unable to attract the science and engineering talent needed for the war and to be unresponsive to military direction, having their own political protectors in Congress. The military, however, was reluctant to turn over all of this work to profit-making organizations. Special not-for-profit organizations, several affiliated with major research universities, were created to help the military design and manage weapon projects. It was a combination of military program managers, not-for-profit technical advisers, and private defense contractors that designed and built America's weapons during the Cold War.

Since the end of the Cold War there has been a tendency to turn even more of this system over to private contractors. Large portions of the government's own technical facilities have been consolidated or closed.[21] Much of the military's equipment repair and technical support work has been privatized. Several of the military-dedicated not-for-profits have converted to private ownership. For example, the infrared laboratory at Michigan has become IRIM International, and parts of MITRE have spun off, including a unit that has entered the commercial software business.[22] Some of the technical work previously reserved for the not-for-profits has been opened to commercial competition. SAIC, a private firm, is now an important systems integrator for military information networks and has a very broad set of consulting and support service contracts with each of the armed services, work that any of a number of not-for-profits used to have as their own. In terms of support, facility ownership, and technical expertise or even management, the weapon development and system coordination business is much less of a public enterprise than it was once.

The trend seems to be in the opposite direction for the social science component of military research. More and more of this work is being done in military facilities. The military's educational network has expanded. There is now an Acquisition University, a Marine Corps University, and an Information Warfare College, among a number of other military institutions of higher education. Located in places like the U.S. Military Academy at West Point; the Naval Post Graduate School in Monterey, California; the Air University in Montgomery, Alabama; and the Air Force Academy in Colorado Springs, Colorado, are military-managed research groups that advise the service chiefs on policy issues. Even when the services seek outside assistance on policy issues, the scope of their search is not wide. Contracts often go to military retirees acting as consultants, service associations, and a handful of companies also led by former officers.[23] Policy research is much more a matter of trust and loyalty than is natural science or engineering research.[24] In science and engineering research the organization's preferences should never influence the results, and in policy research the organization's preferences are the very reason why assistance is sought.

Today's military fights less at home as well as abroad. A tolerance for interservice rivalry has been supplanted by what could be called the religion of "Jointness," the belief that the services should cooperate in operations, force planning, and weapon acquisitions in order both to be a more effective fighting force and to avoid being vulnerable to civilian manipulation. Bickering among themselves, the services are likely to point out flaws in each other's arguments. These professional insights into the limits of a service's doctrine or technology give civilians in the Office of the Secretary of Defense or on congressional staffs leverage in policy discussions.[25] Now, thanks no doubt to Jointness, it is not unusual to see the navy seeking credence for a favored position by sponsoring an endorsing study at RAND, one of the air force's chosen study organizations. What better way to convince somewhat ill-informed outsiders of the virtues of your plans than to have an "opposing" organization offer certification of its wisdom?

BECAUSE the intent in official descriptions is so often to mislead, a bit of cynicism in assessing the contribution of science, social or natural, to warfare should be welcomed. The Cold War fortunately involved little direct fighting and much preparation. The knowledge of natural scientists and engineers was very much needed to develop the weapon systems that would have been used against Soviet forces and were occasionally used against the forces of Soviet allies. But mostly, the Cold War's weapons were edged into obsolescence by new systems that scientists and engineers devised to counter others that Soviets did, could, or might develop. The only certain battles the war's weapon systems had to face and win were the bureaucratic ones determining their development and production funding. Not surprisingly, the military had a need for those among us, often social scientists, who can design the forums, the contest rules, and perhaps even the winning arguments for those crucial battles.

Thanks to military patronage we now have stronger area studies, better knowledge of how large-scale organizations function, and some insight into ways to improve the comprehensiveness of the policy-making process if not the effectiveness of the actual policy outcomes. Social science did enhance national security if only by helping us believe that we were making rational choices among the weapons and strategies, that there was some sanity in the mammoth and dangerous undertaking of confronting globally a nuclear armed opponent for five decades.

Notes

1. Not enough has been written totaling the cost or measuring the impact of the Cold War. A partial approach is to examine how the nation's political leadership tried to keep a balance between an all-consuming war and domestic consumption. Note Irving Bernstein, *Guns or Butter: The Presidency of Lyndon Johnson* (New York: Oxford University Press, 1996). See also Diane B. Kunz, *Guns and Butter: America's Cold War Economic Diplomacy* (New York: Free Press, 1997).
2. See Aaron L. Friedberg, "Why Didn't the United States Become a Garrison State?" *International Security* 16, no. 4 (spring 1992): 109–142.
3. Geoffrey Perret, *And There Is a War to Win* (New York: Random House, 1993). Cf. Martin van Creveld, *Fighting Power: German and U.S. Army Performance, 1939–1945* (Westport, Conn.: Greenwood Press, 1982).
4. Barry Posen, "Military Format Analysis," forthcoming.
5. This history is covered in Harvey M. Sapolsky, *Science and the Navy: The History of the Office of Naval Research* (Princeton, N.J.: Princeton University Press, 1990). For science-navy relations in the pre–World War Two period, a good place to start is William M. McBride, "The Greatest Patron of Science?: The Navy-Academia

Alliance and U.S. Naval Research, 1896–1923," *Journal of Military History* 56 (January 1992): 7–23.

6. Sapolsky, *Science and the Navy*, 29–36. Today the organizational legacy of the scientists' quest for independence lies in the much-praised (usually by scientists themselves) Defense Advanced Research Projects Office (DARPA). For the latest ode to DARPA, see David Malakoff, "Pentagon Agency Thrives on In-Your-Face Science," *Science* 285, no. 3 (September 1999): 1476–1479.
7. Harvey M. Sapolsky, Eugene Gholz, and Allen Kauffman, "Security Lessons from the Cold War," *Foreign Affairs* 78, no. 4 (July/August 1999): 77–89.
8. Sanford Weiner, "Resource Allocation in Basic Research and Organizational Design," *Public Policy* (spring 1972): 227–255.
9. See Daniel Wirls, *Buildup: The Politics of Defense in the Reagan Era* (Ithaca, N.Y.: Cornell University Press, 1992), esp. 133–168; The difficulties in making any estimate of Soviet military expenditures, let alone their cause, is discussed in Noel E. Firth and James H. Noren, *Soviet Defense Spending* (College Station: Texas A & M Press, 1998).
10. Wisla Suraska, *How the Soviet Union Disappeared* (Durham, N.C.: Duke University Press, 1998), 65. Chief of staff of the Soviet armed forces Marshall Nikolai V. Ogarkov's fears of American capabilities and intentions expressed publicly in the early 1980s were an impetus for restructuring the Soviet economy. These reform efforts, it could be argued, led to the unraveling of the Soviet Union. For some of Marshall Ogarkov's views, see Sherry Sontag and Christopher Drew, *Blind Man's Bluff: The Untold Story of American Submarine Espionage* (New York: Public Affairs Press, 1999), 243–244, 275.
11. Herschel Kanter, "Defense Economics: 1776–1983," *Armed Forces and Society* 10, no. 3 (spring 1984): 436.
12. See Peter Buck, "Adjusting to Military Life: The Social Sciences Go to War, 1941–1950," in *Military Enterprise and Technological Change*, ed. Merritt Roe Smith (Cambridge: MIT Press, 1985), 203–252. Another good perspective is provided by David R. Segal, "Measuring the Institutional/Occupational Change Thesis," *Armed Forces and Society* 12, no. 3 (spring 1986): 351–377. Examples of the work in this vein are Hyder Lakhani, "Reenlistment Intentions of Citizen Soldiers in the U.S. Army," *Armed Forces and Society* 22, no. 1 (fall 1995): 117–130; Morris Janowitz, "Consequences of Social Science Research on U.S. Military," *Armed Forces and Society* 8, no. 4 (summer 1982): 507–524.
13. Note, for example, Robert Roehrkasse, "Comparative Staffing and Advancement Patterns: The Air Force and the Private Sector," *Armed Forces and Society* 8, no. 4 (summer 1982): 601–614; William E. Rosenbach and Robert A. Gregory, "Job Attitudes of Commercial and U.S. Air Force Pilots," *Armed Forces and Society* 8, no. 4 (summer 1982): 615–628; Gene I. Rocklin et al., "The Self-Designing High-Reliability Organization: Aircraft Carrier Flight Operations at Sea," *Defense Analysis* 9 (1993): 31–42.
14. Allan A. Needell, *Science, Cold War, and the American State: Lloyd V. Berkner and the Balance of Professional Ideals* (Washington, D.C.: Harwood Academic Publishers/Smithsonian Institution, 2000), ch. 6; Donald L. M. Blackmer, *The MIT Center for International Studies: The Founding Years 1951–1969* (Cambridge: MIT Center for International Studies, 2002), ch. 1.
15. Alain C. Enthoven and K. Wayne Smith, *How Much Is Enough? Shaping the Defense Program, 1961–1969* (New York: Harper and Row, 1971).
16. See Frederick C. Mosher and John E. Harr, *Programming Systems and Foreign Affairs Leadership: An Attempted Innovation* (New York: Oxford University Press, 1970).

17. Arnold Kanter, *Defense Politics: A Budgetary Perspective* (Chicago: University of Chicago Press, 1979). See also Richard P. Nathan, *Social Science in Government* (New York: Basic Books, 1988), esp. ch. 2.
18. James R. Schlesinger, "Uses and Abuses of Systems Analysis," *Survival* (October 1968): 334–342; Stephen P. Rosen, "Systems Analysis and the Quest for Rational Defense," *Public Interest* (summer 1984): 3–17; Aaron Wildavsky, "If Planning Is Everything, Maybe It's Nothing," *Policy Sciences* 4 (1973): 127–153; Thomas E. Anger, ed., *Analysis and National Security Policy* (Alexandria, Va.: Center for Naval Analysis, 1988).
19. Gregory Palmer, *The McNamara Strategy and the Vietnam War: Program Budgeting in the Pentagon, 1960–1968* (Westport, Conn.: Greenwood Press, 1978).
20. One of the key battles was the battle over the direction of military projects by the technical branches of the military. McNamara succeeded in stripping the military of its main technical design institutions, forcing a dependency on industry. Of course, the most-remembered involved the bitter (and continuing) conflict over the bombing strategy in Vietnam. See Richard S. Greeley, "Stringing the McNamara," *Naval History* 11, no. 4 (July/August 1999): 60–66; Larry Cable, "The Operation Was a Success, but the Patient Died: The Air War in Vietnam, 1964–1969," in *An American Dilemma: Vietnam 1964–1973*, ed. Dennis E. Showalter and John G. Albert (Chicago: Imprint, 1993) 109–158; Earl H. Tilford Jr., *Setup: What the Air Force Did in Vietnam and Why* (Montgomery, Ala.: Air University Press, 1991).

    There were indeed some actual Whiz Kids, the popular term at the Pentagon for McNamara's analysts, but also the term applied to the brash young veterans, McNamara among them, who took over at Ford Motor Company to save it from its growing management crisis. John A. Byrne, *The Whiz Kids* (New York: Currency, 1993).
21. Harvey M. Sapolsky and Eugene Gholz, "Private Arsenals: America's Post Cold War Burden," in *Arming the Future: A Defense Industry for the 21st Century*, ed. Ann R. Markusen and Sean S. Costigan (New York: Council on Foreign Relations, 1999), 191–206.
22. This is another understudied and underreported change in defense policy following the end of the Cold War. In essence, a chunk of the public investment in defense capabilities was privatized, sometimes to the direct benefit of the assets managers, with competing private organizations also benefiting from the removal of rivals.
23. The former deputy chair of the Joint Chiefs of Staff, Admiral William Owens, retired to head SAIC, one of the largest of these firms. Secretary Richard Cheney, on leaving office, became the chief executive officer of Halliburton, the owner of Brown and Root, and a large logistics support contractor for the military.
24. So, too, was management once as is discussed in Terrence J. Gough, "Origins of the Army Industrial College: Military-Business Tensions After World War I," *Armed Forces and Society* 17, no. 2 (winter 1991): 259–276. But perhaps with the steady flow of retired officers into the upper ranks of the contractors, the concern about managerial control has relaxed.
25. Harvey M. Sapolsky, "Interservice Rivalry Is the Solution," *Joint Forces Quarterly* (spring 1997): 50–53. Note also Ernest R. May, "Intelligence: Backing Into the Future," *Foreign Affairs* (summer 1993): 63–72.

Chapter 4

A Risk Perceived Is a Risk Indeed

Assessing Risk in Biomedical Research and Health Policy

PHILIP L. FRANA

CONTEMPORARY RISK-ASSESSMENT literature often separates national health promotion policies from national biomedical research policies. Where popular health promotion showcases the enigma of "good health" versus "personal freedom," biomedical research is interpreted as a zero-sum game where ambitious research agendas are constrained by limited public funding. The epistemological bifurcation of biomedical research and health policy, however, neglects an underlying conjunction between the two areas in the contemporary risk assessment-abatement debate. In both cases, individuation and "democratic" decision-making altered our most basic assumptions about the nature of risk in and out of the laboratory. Risk, moreover, now represented negative outcomes to avoid rather than as—on balance, at least—neutral events in the game of life.[1]

Health assessment is a large and growing concern of the federal government. In 1993 the Office of Technology Assessment estimated that total funding for health assessment research exceeded half a billion dollars. This type of research was found suffused throughout most government agencies in the 1990s but remained largely uncoordinated. Federal support for biomedical and health risk assessment research was funneled to more than twelve different agencies. The Centers for Disease Control identified and characterized risks through the Center for Environmental Health, the Center for Infectious Diseases, the Center for Health Promotion and Education, the Center for Prevention Services, and the National Institute of Occupational Safety and

Health. The Food and Drug Administration (FDA) distributed this work, directed toward the goal of consumer protection, to all five of its centers: the Center for Food Safety and Applied Nutrition, the Center for Veterinary Medicine, the Center for Drugs and Biologics, the Center for Devices and Radiological Health, and the National Center for Toxicology Research. Risk assessment was—and remains—also a primary mission of the National Institutes of Health, Environmental Protection Agency, Occupational Safety and Health Administration, and Consumer Product Safety Commission.[2]

Several more or less ephemeral federal representative bodies addressing biomedical and health risk have also come into and gone out of existence. The National Research Council, Food and Drug Administration, Occupational Safety and Health Administration, National Institutes of Health, Consumer Product Safety Commission, and Environmental Protection Agency all inaugurated committees that presented leading but substantially different risk assessment guidelines in the 1970s and 1980s, differing particularly in their sensitivity to potentially cancer-causing compounds and in establishing criteria employed in the testing of chemicals. More recently, the U.S. Congress constituted a Commission on Risk Assessment and Risk Management in 1990 to hear testimony from interest groups with "views on public and environmental health regulatory issues." Testimony continued for three years. In June 1997 the commission released its hefty report arguing that a "chemical-by-chemical" or "risk-by-risk" approach to health found in previous federal guidelines had reached its maximum potential. It recommended that more attention be paid to matters of personal behavior and responsibility to further reduce risks to health.[3]

Much can be learned about the portrayal of risk in these institutions and committees simply by looking at a number of specific policy issues. National biomedical and public health policy regarding xenotransplant procedures and breast cancer screening in older women are two arenas where a particularization of roles and democratization of process generated amorphous standards toward and ambivalent responses to risk in contemporary America. The singularly American preoccupation with obesity and dietary fat, where accountability lay with each informed individual, is another. Experimental medical procedures and drug trials, where otherwise terminal patients became motivated to personally research medical trials and adopt a "consumerist perspective" in negotiations with physicians before enrolling or giving consent, also reveal systemic fragmentation of authority. At another level, the transformation of health insurance policies from "community" to "experience" rating matched a deeper societal transformation from associational to individualist form. Each of these issues will be touched upon here.

In the last week of June 1995 a group of over two hundred immunologists, surgeons, medical ethicists, attorneys, administrators, historians, humanists, sociologists, and representatives of corporate and nonprofit organizations, special interest groups, and the interested public gathered in the Crystal Ballroom of the Bethesda Hyatt Regency Hotel in Maryland. They were there to participate in a crucial and timely Institute of Medicine–sponsored Xenograph Transplantation Workshop, devoted to examining the possibility of lifting a voluntary moratorium on clinical trials involving human-animal organ exchange or, alternatively, of placing further limits on cell and tissue exchange trials planned or already in progress.[4]

Chimerical combinations of human and beast have long inspired spectacular visions in the Western mind, from the Greek bull-man Minotaur and the half-horse, half-human centaur to the modern mutated human Wolverine of X-Men comic books and the Flukeman, king of the sewers in the *X-Files* universe. Few, if any, participants in the xenotransplant debate worried that health scientists and policy-makers desired or even possessed the tools to breathe real life into such hybrids. Still, many articulated urgent, even apocalyptic, concerns about the potential personal and public risks of human-animal transplantation.

Xenotransplantation promised new treatments, perhaps even cures, for some of humankind's most intractable diseases, diseases like cancer, AIDS, diabetes, and Parkinson's. Paralleling hopes, however, were concerns that human-animal exchanges might introduce previously unidentified animal infections into human populations. Some workshop participants flatly refused to endorse future experiments involving the transplantation of animal cells, tissue, or organs into human subjects because, as the final report noted, the degree of risk "is unequivocally greater than zero." Others thought the benefits greatly outweighed the risks and that those risks were manageable.[5]

The executive summary of the Xenograph Transplantation Workshop approved human-animal tissue exchange "when the science base for specific types of xenotransplants is judged sufficient and the appropriate safeguards are in place." It suggested that the Department of Health and Human Services (HHS) establish a xenotransplantation advisory committee to create guidelines but not directly regulate research. Most specifically, it recommended coordination among xenotransplant researchers to establish sources of pathogen-free animals, tissue banks of humans and animals involved in experiments, registries of xenotransplant patients, and, most Orwellian of all, lifetime surveillance of patients and their closest contacts for any signs of emerging diseases.[6]

The recommendations of the Executive Committee of the Xenograph

Transplantation Workshop were tested almost immediately. In mid-July the Food and Drug Administration, overturning its own ruling against human-animal organ xenografts made in late April of that same year, approved the transplant of purified and AIDS-immune baboon bone marrow into a patient with advanced AIDS in San Francisco. The experiment, helmed by transplant surgeon Suzanne Ilstad of the University of Pittsburgh, aimed to restore the patient's ability to manufacture T cells, which attack HIV–1.[7]

On December 13, 1995, a University of California–San Francisco surgeon performed the xenotransplant procedure according to Ilstad's protocols on thirty-eight-year-old Jeff Getty, injecting millions of baboon marrow cells into his body. The FDA's expert committee offered Getty slim odds for success but was moved by the pleas of AIDS activists in agreeing to this test but no others for the time being. AIDS activists, galvanized by hope, paid for Getty's bone marrow transplant operation, a sum that exceeded $40,000.[8]

Getty's transplant sparked a very public debate echoing the more esoteric one facilitated by the Institute of Medicine's Xenograph Transplantation Workshop. Many experts worried about the public health risks of the xenograft despite Getty's pledge—if he survived—to practice safe sex and submit to the taking of regular blood and tissue samples. Virologist Jonathan S. Allan of the Southwest Foundation for Biomedical Research in San Antonio, who abstained from the FDA's vote of approval, summarized the view of many plainly: "If you don't want to risk the public health, then don't do it."[9]

Other experts questioned the validity of Ilstad's scientific claims. Stephen Rose, director of genetics and transplants at the National Institute of Allergy and Infectious Diseases, noted with concern that from "what I've seen of the data, I don't see where this warrants going ahead with a human clinical trial." He complained in particular that there was no independent proof that the facilitator cells Ilstad believed helped acclimatize baboon marrow to humans and which helped prevent rejection actually existed. David Nathan, president of Boston's Dana-Farber Cancer Institute agreed: "There's a very healthy skepticism about these findings." Others characterized Ilstad's experiment as a blunt and indiscriminate tool against a problem requiring finer instruments.[10]

Ilstad responded to these complaints by arguing that there was "an art" involved in the isolation of facilitator cells in baboon marrow and that the merits of her pending patent application for the procedure were actually strengthened by the great difficulty in finding them. "If it was something so obvious," she countered, "it couldn't be patented." Martin Delaney, director of the AIDS advocacy group Project Inform, backed up Ilstad's claims repeatedly in the press, insisting that naysayers who feared potentially revolutionary therapies quit practicing "comic-book hysteria" and "get a life."[11]

Controversy also swirled around Getty's progress following his xenograft operation. By late January Getty's doctors concluded that most, if not all, of the baboon cells transplanted had died or failed to replicate. Yet Getty himself appeared to grow stronger. In February his weight stabilized, and his asthma and seborrheal eczema problems disappeared.[12]

The concerns voiced at the Xenograph Transplantation Workshop and the Getty xenograft procedure were by no means unique. They were, indeed, manifestly representative of the reactive climate surrounding American biomedical and public health policy generally. It was an omnipresent "fact" in contemporary America that progress in science and technology ratcheted up the risks of failure to potentially catastrophic levels, unleashing "irreversible" hazards and "cascading" chain-reaction failures. Xenotransplant research was identified as one of these hazardous health-related advances, but many more were recognized. Antibiotics, the great lifesaver of the last half century, were now condemned by some as overprescribed, responsible for flesh-eating, drug-resistant bacterial infections. Chlorofluorocarbons (CFCs), used as an element in air-conditioning, which saved the lives of so many elderly Americans during heat waves and made much of the South livable, were now found to blame for opening a gaping hole in the earth's protective ozone layer. Global manufacture and distribution of milk and meat, vitamin and protein sources for millions in this country, now threatened to saturate the public with artificial growth hormones and unleash "mad cow" and E. coli epidemics.[13]

But one needed only to take a brief excursion into contemporary etymology to detect an unfolding siege mentality. The word "risk" no longer evoked neutrality or even restraint. Risk only a few decades ago merely identified the ever-present factors shaping a whole constellation of hereditary and environmental characteristics circumscribing human existence. Risk represented the odds attached to personal physical and mental health as the inherent price of doing business—that business being living life productively, zestfully, and without pervasive worry. In many ways, as Richard Meehan has pointed out in the oft-quoted *The Atom and the Fault* (1984): "The need for or even existence of a higher standard of safety never occurred to men who had survived World War II, smoked a pack of Camels a day, and were faced with a hostile Soviet adversary armed with ICBMs. The world was a dangerous place and life was short."[14]

Risk once equated with contingency, but in contemporary society risks were dangers to be avoided by rational individuals whenever possible. Government experts encouraged Americans to deprive themselves of the simple rewards of affluence, of saturated fats, cigarettes, artificial sweeteners, television viewing, and Internet browsing, in exchange for improvements in the personal odds against obesity, heart attack, hypertension, mental disease, cancer, and untimely death. At no time in American history had the notion of risk become so bound to cost that the words could now be used interchangeably. "Cost-benefit" and "risk-benefit" referred, in most cases, to the same kind of analysis. And the interchangeability of cost-benefit and risk-benefit rested on a significant premise: that econometric tools were appropriate ways to capture risk in its natural and human-made environments. Simply put, risks in contemporary America were intimately associated with and understood in terms of their money value.[15]

That simple change in definition had enormous implications for business law alone. It was no longer enough to have good intentions behind product development and distribution in order to avoid litigation. Industry had a new worry beyond the intentional tort, the deliberate infliction of harm, and beyond negligence, harm by ineptitude. Now it had to protect itself from mistakes in calculating the costs and benefits of its products to avoid product liability suits. Manufacturers brought to every market decision new worries about design, use, and warning labels to protect themselves from consumers who might unwittingly harm themselves through inappropriate use of products. In hospitals, risk-benefit analysis convinced many childbirth specialists to choose cesarean delivery over natural childbirth as a check on malpractice claims and insurance premiums.[16]

Contemporary preoccupation with risk as a negative phenomenon also left an indelible impression upon personal behavior. Indisputably, the newer assessment of risk reshaped conceptions about the acceptability of risk but at the same time did not reduce in any appreciable way the personal or social risks of life itself. The cacophony of hazard warnings broadcast into an increasingly harried world did not seem to make life any less dangerous or any more pleasurable. Risk managers and policy-shapers, considered generically as those serving or entrusted as the guardians of our good health, in their efforts to reduce risks did not change perceptions about our national life for the better. Instead, they encouraged regularized, even institutionalized, self-flagellating bouts of personal guilt. Self-discipline in a society where denial was rare was a scarce commodity, and Americans felt worse about themselves even as they were told—sometimes in excruciating detail and sometimes with airy platitudes—how to feel better.[17]

No wonder some Gen-Xers, prideful Mountain Dew–swilling fatalists, were driven to ski down impossible Alaskan slopes, ride titanic Hawaiian waves, and run across burning California deserts. Otherwise "risk-averse" individuals ate, drank, and slept their way to happiness only to be tormented by a conscience reminding them that this path led inevitably to personal dissolution, disease, and destruction. Ironically, perhaps, some Americans nostalgically embraced risk taking as a primary outlet for emotional distress caused by risk aversion. We became a binary people, alternately bingeing and purging ourselves with little thought for moderation, a proposition good enough for Hippocrates, Galen, and the president of the American Dietetic Association but apparently not satisfactory to the rest of us.[18]

Accountability in matters of health, even with health policy made in far away Washington, lay with the ordinary American. Independent achievements in reforming personal behavior became what mattered, the sum making the whole in moving the nation across the bridge to the next century. In essence, public health disappeared, and personal health took its place. This impulse worked at some of the most fundamental levels of biomedical and health research practice. Potential patients—even the mentally impaired—were encouraged by experts and laypeople to bone up on the tests, drugs, and operations to which they might be subjected and to insist upon consensual relationships with members of the "medical establishment."[19]

This was true, too, of experimental procedures where otherwise terminal patients were motivated to research medical trials before enrolling. The Biological Therapy Institute, a for-profit medical research corporation, went so far as to tout its efforts to re-create "patients as research partners." Web sites proliferated to enlarge public access to experimental trials and educate the infirm. CenterWatch hosted a listed service of trials ranging from oncology to nephrology to rheumatology to gastroenterology. The National Cancer Institute (NCI) offered a similar page offering access to participation in its own cancer trials.[20]

The process also worked in reverse. Individual transgressions subtracted from the overall health of the nation. Risk-takers also made risk-makers. As John Knowles, former president of the Rockefeller Foundation, put it, paraphrasing Pogo, "We have met the enemy and it is us!" Individuals, with the entire onus for self-education and wise decision-making (symbolized by the new catchword: "empowerment") in matters of health upon them, were confronted daily with a fusillade of public service messages from a panoply of health risk managers. Self-education was no easy proposition and for some grew to unhealthy proportions. "Do-it-yourself" health assessment handbooks, magazine checklists, and home diagnostic tests—from pregnancy tests to devices checking for HIV and yeast infections, cholesterol levels, and illegal drugs in a teenager's system—proliferated.[21]

The media also supplied a bewildering assortment of outlets for improving individual health. HealthScout touted its online service supplying "personalized health news for you and your family." *In Touch Magazine* gave people advice about the relationship between personal health, creativity, professional development, and world ecology. The Chedoke Peer Support group published a monthly newsletter covering the health of amputees. Staid Harvard Health Publications offered no less than five personal health-related publications, the *Harvard Health Letter*, *Harvard Heart Letter*, *Harvard Mental Health Letter*, the *Harvard Women's Health Watch*, and the *Harvard Men's Health Watch*.

Coupled to this great and rapidly enlarging body of popular health literature were regular press reports abstracting respected medical journals for mass consumption. Stories selected from *JAMA* and the *New England Journal of Medicine* often reached the public ear with the morning coffee, well before falling into the hands of subscribing physicians. Deflation inevitably supplanted elation over the latest reported advance as the overriding goal of instantaneous access to information stacked the playing field against careful scientific scrutiny. Scientific methodology—particularly the doctrine of falsifiability with its core assumption of nullification as the road to truth—became diminished in a sound-bite world where competition for viewer, listener, and reader attention was intense. Hailing the triumphs of each individual advance in medical science—and undeniably there had been many—subverted the established methodology of science used in the generation of each advance, for only when considered as a whole raft of scientific studies did the Popperian program make sense.

Decision-making, ultimately, was to be shared between the public and the experts, with no clear leader and no clear subordinate. Patients adopted a "consumerist perspective" whereby they became "clients" bargaining with their physicians and following institutional guidelines like the multitudinous forms of the "Patient's Bill of Rights." Maintaining and improving health were dealt with much the same way individuals handled other aspects of their lives. Patients labored under the assumption that their care providers, like their golf pros and brokers, did not have nearly as much riding on the outcome as they did. The risk that patients assumed went largely unshared.[22]

A case in point was the health care consumerism built into the Federal Employee Health Benefits Program, where a highly fragmented system of selecting health care products evolved as a larger and larger share of costs was borne by the employees themselves. Every year federal government employees selected their own health plans à la carte from a smorgasbord of health maintenance organizations, patient provider organizations, and indemnity and point of service plans. Described by employers as a "highly competitive" solution to the problem of health care benefit distribution, employees were assisted in their decision-making by 800 numbers, numerous guides to alternative plans, and comparison charts, but in the final analysis the clients lived with the risks associated with their own choices.[23]

The deinstitutionalization of decision-making, though, made it difficult for government and its experts to modify behavior and expectations of behavior in contemporary society. Government, in fact, had in many ways abetted the individuation of risk selection and abatement by making their taking a "voluntary" matter. For example, the historian Allan M. Brandt and others note that government warning labels placed on cartons of cigarettes and in cigarette advertising have shifted the burden of "responsibility for smoking and its risks from the industry to the individual smoker." Proper assay and mass presentation of particular "risk factors" absolved industry of the consequences because the individual had become educated against smoking but did so anyway. The same held for government drug and AIDS prevention programs. Each was "caused by a moral failure of the individual." Constructing these diseases as the results of lifestyle choices absolved government because the individual was ostensibly educated against partaking in these risky practices. Personal conduct was the problem and not—for once—government.[24]

Government also succored individuation with overspecialization in seeking to disengage risk abatement from risk assessment. Those who actively engaged in research on carcinogenic substances, for example, were divorced from the implementers of cancer prevention programs. Thus those who devoted their time to basic research in the biomedical and health sciences—those most intimately aware of laboratory and clinical manifestations of risk—were no longer directly responsible for supervising day-to-day abatement activities. Despite complaints by researchers, policy-makers, and administrators, an institutionalized particularization of roles divided risk management into researchers and technicians.

Expertise in the early and mid-twentieth century drove the decision-making process. Professionals promulgated health and quality-of-life standards and gathered together enforcement agencies. Experts used their mastery of an esoteric body of knowledge derived from (often long) periods of formal education and training to legitimate their positions as ultimate arbiters of the nation's health. No longer. Biomedical and health policy in contemporary society was made ideally from the "bottom up" through "consensus building" and only sparingly and ineffectively from the "top down."[25]

Further, experts now generated so much advice, so much that was reported as controversial, and so much that was contradicted by other studies that informed Americans no longer knew where to turn or whom to trust. For a long time athletes, particularly ball players, worried about minimizing tears to their rotator cuffs, a common injury in sport. Now, however, they were told to isolate and exercise the scapular stabilizer and measure its imbalance with the controversial Lateral Scapular Slide Test. Athletes also endured a long string of tests designed to quantify body fat. Skin calipers, submergence, formulas, bioelectrical impedance, and dual energy X-ray absorptiometry tests were all touted at various times and places as the best available measure.[26]

The general public endured conflicting expertise concerning weight management, too. Consider the case of obesity.[27] Merely identifying the obese according to the esoteric language used in professional scientific discourse proved difficult. As the late Ancel Keys of the University of Minnesota Laboratory of Physiological Hygiene put it fifty years ago, "I suppose all educated people so understood the word until the fashion of measurement saddled us with the habit of trying to decide on the existence of obesity by the use of scales and somebody's table of how much we should weigh." Tables of "ideal weights" can be traced back to efforts by insurance companies in the first decades of the twentieth century to establish desirable weight targets for men and women. Tables of ideal weight had serious limitations, foremost among them that they were normative measures in the sense that they did not always reflect average weights found in actual human populations, nor did they allow for direct comparison of actual adipose tissue (body fat) or skeletal "frame size" within or between weight classes. Tables measured only what was average, not what was normal. Ideal body weights, even corrected for height and age, for instance, tended to identify strapping football players as obese even when by health and aesthetics they were not.[28]

Already by the 1940s the medical profession had developed several devices for estimating "total body fat." Skinfold calipers, soft tissue roentgenograms (X-ray silhouettes), densitometry (immersion in water), and antipyrine injection (or the "total dilution method," measuring the rate of diffusion of a chemical in the nonfat mass of the body) all were used to estimate subcutaneous fat percentages independent of body weight. These measurements figured prominently in tables and equations used in physiological experiments, but difficulty in making consistent estimates of body fat percentages in laboratory subjects limited their utility. Popular health campaigns continued to promote weight charts, today represented as the Body Mass Index (BMI), calculated by dividing body weight by the square of height.[29]

Before the 1950s experts explained most forms of obesity, at least in part, in terms of a wider dysfunction in community and familial patterns. That is to say, family ancestry and social disorganization triggered human obesity. Of course, these experts understood that glandular imbalances and overeating existed, but such factors did not necessarily explain the root cause of the dysfunction. An unbalanced personality formed from adverse familial or cohort assessments of weight and shape could easily lay the groundwork for an obese personality. Regardless, most authorities tended to agree that overweight and obese people suffered proportionally more ill health in the form of circulatory disease, nephritis, cirrhosis, diabetes, and a long list of other somatic and mental afflictions than those of average weight.

This associational or familial view of obesity continued in strength into the early 1950s and then crumbled. "Fat pad" or skinfold testing of Micronesian boys and girls, Boston children, and U.S. Army personnel by Harvard's Peabody Museum, the Office of Naval Research, the Pacific Science Board of the National Research Council, the Departments of Nutrition and Anthropology of Boston's Forsyth Dental Infirmary, and the Charles H. Hood Dairy Foundation, for instance, revealed no correlation between obesity and nativity, disturbed communities or families, or maternal morphology.[30]

Despite a long history of statistical analyses linking overweight people to excess mortality, some experts began to attack the premise that all obesity was harmful. It might even be "fit to be fat." Led by the vocal critic Paul Ernsberger, a biomedical researcher in the Cornell University Medical School, these dissenters from conventional wisdom argued that substantial evidence existed showing that obese people had life expectancies equal to that of "normal" people. Moreover, obese people also had reduced risks for certain afflictions—compensating for any elevated risks. Ernsberger noted that obese Americans suffer—among other things—less cancer and renovascular hypertension, fewer infectious diseases, and lower rates of suicide. Less radically perhaps, Rubin Andres of the Institute on Aging began arguing that progressive weight gain throughout life produced negligible increases in overall mortality rates.[31]

Other experts, many self-described feminist scholars, argued that overweight and obesity existed only as aesthetic and moral problems skewing human perception of a healthy "body image." Fat people were, in sum, "victims" of thin people who unhealthily obsessed about their own weights. Obesity represented not so much a public health problem but a "social embarrassment." As evidence, these detractors pointed out generational differences in the average size, weight, and shape of beautiful women. They claimed that to be fat was actually to be "exceptional, expansive, strong, warm, generous, [and] nurturing." They did not advocate weight loss in most cases and emphasized the health risks of "yo-yo" dieting or "weight cycling" for the obese.[32]

Most experts in contemporary America, however, clung to the mantra "overweight is risking fate."[33] For a majority of physiologists, nutritionists, and physicians who agreed about the overwhelmingly negative health effects of overweight and obesity, personal responsibility outstripped family or community responsibility for deteriorated condition. Obesity for them was not only ugly; it was a sickness stemming from, besides glandular difficulties, feelings of personal inferiority, sexual inadequacy, and lack of willpower. Children overate as a form of personal rebellion against their parents. Adults overate to derive consolation from others and to defy social expectations. Fat became a personal cross upon which the sins of society unfairly attached themselves. Some of the experts in this group battled overindulgence by seeking to "pep up" obese people, hoping to instill self-control by building up a "sense of pride and self-esteem among overweights." Discipline always came from within.[34]

The redefinition of federal obesity guidelines in 1998 did little to ease the worries of the body conscious and the corpulent. Revised standards for overweight and obesity found in the famous *Healthy People 2000* report (which cribbed standards with few revisions from standards established in the 1940s by the Metropolitan Life Insurance Company) by the National Heart, Lung, and Blood Institute along with the National Institute of Diabetes and Digestive and Kidney Diseases generated considerable professional and popular interest. It reduced the threshold for the overweight and obese classifications substantially to "reflect the enormous advances in medical science" by shifting lower BMI scores into "at-risk" and "high-risk" categories. Many people who before were considered normal under the old BMI charts now found themselves overweight, and many among the overweight suddenly found themselves obese. In 2002 a breathtaking 65 percent of adults combined fell into the overweight and obese categories.[35]

Science contributed to American gluttony with such advances as "low fat" ice cream and olestra. Due in no small part to marvelous advances in food technology, Americans in 1990 consumed fewer calories from fat than they did in 1960 but 14 percent more total calories. The disconnect between new and restrictive federal BMI charts and the consumption of mass quantities of food and beverages resulted mainly in frustration, guilt, and, of course, more overeating.[36]

The history of breast cancer prevention is also a revealing case in point. Here, expert ambiguity by default and design transformed millions of women in their forties into the sole arbiters of risk regarding their own breast health. The American Cancer Society (ACS) estimates that one in nine women will be diagnosed with breast cancer sometime in her lifetime; one in thirteen women will die of the disease. During the decade of the 1980s more than 400,000 women died from breast cancer in the United States, and today approximately 175,000 new cases are identified each year.[37]

According to the ACS, radiological advances, periodic self-examination, and regular physician palpitation have increased five-year survival rates in women with early and localized breast cancers from 78 percent in the 1940s to 91 percent in the 1990s. Despite this progress, the contemporary record with regard to breast cancer screening is incontrovertibly a checkered one. A widely held apperception of shoddy standards and lack of uniformity in the use of machines by poorly trained technicians, and of specialists unevenly interpreting results, have plagued mammography into the 1990s. Indeed, in a 1992 hearing before the Subcommittee on Aging of the Committee on Labor and Human Resources of the U.S. Senate, Dr. Juleann Gandara, chief of mammography in the Department of Radiology at University of Washington, described many screening clinics as "doc-in the box setups . . . where you put out your sign and, you know, you can have your Pap smear, your mammogram, you can have your hair curled and your nails done."[38]

The tamoxifen debate, which the Human Resources and Intergovernmental Relations Committee of the Committee on Government Operations of the U.S. House of Representatives situated glumly under the header "Are Healthy Women Put at Risk by Federally Funded Research?" in a 1992 report, also interfered substantially with efforts to maintain a fragile professional and public climate of optimism about mammography. Tamoxifen is an approved pharmaceutical—in use in pill form for more than three decades against late-stage breast cancers—that changes the activity of estrogen circulating in the body and keeps cancer cells at bay. Tamoxifen has also been shown effective in suppressing the spread of breast cancers in the earliest stage and as an appropriate therapy for cancer-free women identified as "high risk." Over four million women had used tamoxifen to suppress cancer formation by 1990, a total of 41,000 woman-years of treatment.[39]

By the early 1990s, however, an accumulating body of evidence led some oncologists, public health authorities, biomedical ethicists, and consumer advocates to label tamoxifen as a potentially life-threatening drug. They were troubled, variously, by statistics suggestive of elevated risks for thrombosis, phlebitis, uterine cancer, premature onset of menopause, cataract formation, memory loss, and liver cancer. Side effects of the drug, including hot flashes, vaginal bleeding, dizziness, fatigue, headaches, vomiting, and depression, also concerned biomedical and public health investigators and administrators.[40]

In 1992 the House Committee on Government Operations became worried enough to launch its own hearings on tamoxifen, focusing especially on a recently inaugurated $68 million double-blind study of tamoxifen prophylaxis in 16,000 women sponsored by the NCI. Comparing the "mixed message" emanating from this NCI Breast Cancer Prevention Trial to the "tarnished record" of the Public Health Service's Tuskegee syphilis study, a parade of legislators and other invited speakers systematically peppered expert witnesses with pointed questions about this "DES of the nineties."[41]

Many believed tamoxifen's "absolute risks of harm" outweighed any apparent benefits. Adriane Fugh-Berman, medical adviser to the National

Women's Health Network, announced to the crowd that "tamoxifen is too toxic for use in healthy women" and cited a Swedish study suggesting that tumors not suppressed by the drug became highly malignant. Michael W. DeGregorio, professor of oncology at the University of Texas Health Science Center, addressed his comments to the "potential for developing tamoxifen resistance" among the drug's users.[42]

Many speakers also complained about the protocols. Arthur Caplan, director of the Center for Biomedical Ethics at the University of Minnesota, complained about the design of the NCI's tamoxifen study. "There is some evidence that inaccurate, incomplete, or in some situations, incomprehensible, information may have been or is now continuing to be provided to women" regarding the "significant risks associated with tamoxifen," he argued. Adriane Fugh-Berman lamented the fact that risks had been "minimized" or "pooh-poohed" by researchers. Helen Rodriguez-Trias, president elect of the American Public Health Association, voiced concerns about improperly obtained informed consent. "It is quite possible, in the informed consent process," she testified, "to totally bias the information by being somewhat authoritarian, by withholding information or even by body language that implies that you think the questions someone asks are not pertinent." Rodriguez-Trias also expressed her unease that only 4 percent of NCI trial participants were minorities, thereby making the discoveries irrelevant to many.[43]

Directors overseeing the NCI trial viewed these criticisms as unjustified. Bernadine Healy, director of the National Institutes of Health, defended the tamoxifen trial as "well grounded in science" and fulfilling "our mission to advance human health through science." She dismissed criticism of the protocols, describing the process as "well designed" and "tightly and continuously monitored." Further, Healy decried the "sensationalist statements" against tamoxifen that had appeared in media reports and expressed her disappointment that many of the NCI trial's defenders were not present to testify.[44]

Peter Greenwald, director of the Division of Cancer Prevention and Control of the NCI, in his testimony defended the tamoxifen trial as upholding the best principles of science for the public good, as a "gold standard of evidence for prevention research." He described the "model informed consent form" used in selecting participants for the trial as "fully up to date" and always open to modification when conditions dictated. Further, Greenwald disagreed with statements made about potential tamoxifen resistance. "We know of no research data whatever that shows that tamoxifen may increase the rate of breast cancer," he noted. Carl Peck, director of the Center for Drug Evaluation and Research at the Food and Drug Administration, also sided with Healy and Greenwald, dismissing critics who contended that the drug involved too high a risk. "Tamoxifen is not risk free," he responded under oath. "There is no such drug, but risks associated with this drug must be compared with its potential benefits for women at risk of breast cancer."[45]

The significance of the risks associated with tamoxifen, plainly, involved more than a simple statistical summary. Attendees generally accepted the estimate of a 2 percent or less chance of contracting breast cancer over the course of the five-year NCI tamoxifen trial but disagreed in their interpretation of what that percentage really meant. Ethicist Arthur Caplan, responding to a question regarding the estimate in the hearing, noted, "the difficulty with respect to a small risk, a 1.7 percent chance, is that different people are going to view it differently. As we know, there are some women who see themselves as at extreme enough risk to be involved in prophylactic mastectomy to avoid breast cancer, so when we are into the realm of taking drugs other [*sic*] may see this as more trivial."[46]

A mammogram controversy following closely on the heels of the tamoxifen debate further ratcheted up professional and public anxiety over the federal government's breast cancer research and public health policies. Since 1977 the NCI and the ACS had issued periodic guidelines recommending regular mammograms for women, particularly women "at risk" for breast cancer because of family history or genetic predisposition. In 1977 the NCI and the ACS recommended mammograms only for those women who had mothers or siblings with breast cancer. In 1980 the ACS modified its guidelines, recommending a "base line" mammogram for women aged thirty-five to forty and periodic consultation between all female patients under fifty and their physicians about the need for further mammograms. Three years later the ACS modified its guidelines again, recommending that all women between forty and forty-nine get a mammogram every year or two. In 1987 the NCI followed suit, recommending regular mammograms for all women over age forty. These guidelines were subsequently confirmed as official guidelines by a conference of national cancer organizations. Very visible public advertising campaigns and breast cancer detection demonstration projects followed, sponsored by nonprofits and the mammography industry itself. They magnified and redoubled their efforts in the early 1990s with messages like "Breast Cancer Is Curable, If Detected Early Enough" and "Don't Wait! The NCI Recommends That Once You Turn 40, Have a Mammogram Every One or Two Years."[47]

In the waning days of 1993, however, the NCI, overruling its own autonomous advisory board, backpedaled in dramatic fashion. It announced that the science base was insufficient to justify continued support of its own mammography guideline for women between age forty and forty-nine. This included women under fifty who had close relatives with breast cancer. The NCI continued to recommend regular mammograms for women fifty and older. "Basically, it's a statement of what works and what doesn't work," explained Larry Kessler, chief of applied research at the NCI. Edward Sondik, deputy director of the Division of Cancer Prevention and Control agreed. "We felt the most appropriate position for us was to state the facts, the science-based facts as they are. This is consistent with what the National Cancer Institute is all about," he insisted. "We are a science-based agency." The NCI also announced that it wanted to get out of the "guideline" business altogether as it was a research organization and not a public policy organization. Never mind that the latest revelation appeared to be a guideline, too.[48]

The NCI and other federally financed breast cancer centers offered many explanations why regular mammograms seemed useless in women under fifty. Some experts offered the hypothesis that the dense composition of young women's breasts made current mammographic technology and techniques useless. Others suggested that the small precancerous lesions found in women in their forties often could be adequately treated when they reached their fifties. Some experts also proposed that the natural ability of a younger woman's body to make cancers go away might preclude the need for mammograms in women aged forty to forty-nine. Still others thought that unnecessary biopsies and surgeries in women under fifty created scar tissue that hindered effective mammograms later on.[49]

NCI supporters also noted the potential risks of repeated exposure to ionizing radiation and the psychological toll of regular mammograms in women under fifty. "If you take 1,000 women at age 40 and screen them every year until they are 50, 700 will have been told something wasn't quite right," explained Suzanne Fletcher, coeditor of the *Annals of Internal Medicine*. Over 97 percent of these 700 will turn out to be false-positives. Other researchers noticed that almost half of those experiencing false alarms suffered heightened subsequent anxiety, and fully a quarter reported an altered mood state.[50]

Yet statistical evidence and even basic biology seemed to contradict the NCI's recommendation. Roughly 20 percent of all cancers of the breast occurred in women under fifty, as well as 30 to 40 percent of all the woman-years lost to the disease. "There are no data supporting screening, but on the other hand we see cancers in women under 50," noted radiologist Robert McLeland at the University of North Carolina. "Are we supposed to wait until those cancers become palpable?" Artificial or arbitrary division of women into cohorts representing the fifth and sixth decades of life also generated doubts. This "dichotomous grouping" led "to the incorrect decision that the age of 50 is a significant break point when it is not," argued Daniel B. Kopans, associate professor of radiology at the Harvard Medical School.[51]

Many experts agreed with McLeland in arguing that even if little science existed supporting regular mammograms in younger women, the conservative and prudent path ought to be the opposite of the NCI's decision. Clark Heath, vice president for epidemiology and statistics at the ACS, believed that the "unclear" results generated by breast cancer science justified continued support for the older guideline.

The NCI's decision to eliminate its mammography guideline for women under fifty ushered in a prolonged period of confusion and boisterous recriminations, which drew the attention of congressional investigators. Larry Kessler's ambiguous statements at the late 1993 NCI press conference in particular caused much hand-wringing. Getting a mammogram, Kessler had said, was "a decision that only the woman could make."[52] This struck many as contrary to the spirit of the announcement in the first place. Illustrative of the climate of confusion following the NCI's decision was an "Editorial Notebook" column by Joyce Purnick published in the *New York Times*. Wrote Purnick,

> The woman, a friend, is a bright, well-read feminist in her 50s, but she refuses to get mammograms. She revealed this one day as we discussed the controversy stirred up by the National Cancer Institute when it dropped its recommendation. . . . With a triumphant I-told-you-so look, my friend points to the NCI's decision as proof she was right all along. "See what you're saying now?" she asks. "That the radiation from mammograms is dangerous." Huh? Where did that come from? Despite the controversy, scientists emphasize that radiation risk is minimal. "No one is saying there's anything dangerous about mammograms," says Dr. Peter Pressman, a breast cancer surgeon. "Radiation hazard hasn't even come up." . . . This unfounded fear is a most destructive side effect of the controversy. . . . [If] women now convince themselves that the Government's chief cancer research agency is worried about the safety of mammograms, how many cancers will spread and kill?[53]

While Purnick was mostly right in reporting that radiation risk did not appear in the text of the first NCI press release, she was misled in assuming that it was unimportant to the controversy generally. Indeed, long-standing fears of ionizing radiation rapidly reentered the equation as scientists began interpreting the NCI's statements.[54]

On January 13, 1994, the Human Resources and Intergovernmental Relations Subcommittee of the House Committee on Government Operations redeployed to investigate the way the NCI had revised its mammogram guidelines. It demanded that the NCI turn over all documentation regarding its recent decision. Believing that information was still being withheld three weeks later, the subcommittee fired off another letter to the NCI to turn over internal memoranda by February 4. Still, executive minutes and "talking points" did not find their way into the hands of subcommittee members until March 7, the day before hearings opened in Washington.[55]

Dispute over the extent to which the NCI's directors turned over material in a timely and responsible fashion made for an inauspicious beginning to what quickly devolved into a veritable tongue-lashing of the organization and by implication all federally sponsored biomedical and health research. Democrat Ed Towns, an eleven-year veteran of New York's Tenth Congressional District, opened hearings of the NCI's revision of its mammography guidelines with a charge that fueled much of the debate: "What NCI has done, is to unravel the last 7 years of intense campaigning and millions of dollars spent on educating women in their forties about the importance of mammography."[56]

Towns distributed blame for the ill-conceived NCI turnabout widely but focused his ire on the organization's overreliance on science and its cost-saving motives. "Radical retrenchment" by "academics and M.D.'s with little medical experience with breast cancer research," he complained, had "set in motion an agenda to impose science over medicine and change the guidelines." This agenda, he continued, left NCI researchers free to believe that "only the scientific trials dictate the truth." This was not what taxpaying citizens demanded of the NCI, Towns argued. "We cannot make life and death decisions based solely on controversial, randomized clinical trials," he said. "This is not just a matter of pure science, but also of medicine, policy, and the experience of thousands of women in this country."[57]

Towns also criticized the NCI for allowing the cost-saving strategies of health care reform to dictate breast cancer policy. NCI reversal, he said, "forms the best pretext for insurance companies not to cover mammograms," a potential savings of $1 billion each year. "It is frightening," noted Towns, "the continued implication that, in fact, the decision to change the guidelines to not recommend mammography before the age of 50 was actually dollar driven and not medically driven, and that the implied purpose was to make anticipated figures, intended to be spent on medical care, fit the medical requirements."[58]

Criticisms of the NCI by women's caucus Representatives Pat Schroeder, Louise M. Slaughter, and Olympia J. Snowe and by former representative Mary Rose Oakar all followed Towns's opening salvos. Schroeder lambasted the NCI in her remarks, declaring, "This is absolute nonsense and we won't stand for this. . . . I get very angry, I mean, are not 13,500 women's lives enough?" Schroeder further attacked the NCI for supporting a biased, male-dominated federal research agenda. The recommendation, she explained, "fits in exactly with what the caucus has seen with how the Federal Government has treated women's health all across the board." Breast cancer researchers, she argued, "didn't even use female rats over at NIH, which is kind of extreme," and even wasted time on "breast cancer studies in men, which I think is a little unique."[59]

Snowe, in her prepared statement, complained that the NCI's motivation in reversing its recommendation was mainly economic. "It is less expensive not to routinely screen women in their forties," she contended. Yet "instead of utilizing that ounce of prevention at their disposal, NCI has opted out of good sense and good science." Former representative Oakar agreed: "Everybody said, 'Well, it costs so much.' But the savings that you make when you catch breast cancer in stage one or two—it costs $10,000 or less to treat that woman. It costs $165,000 or more if she gets an advanced breast cancer."[60]

Sam Broder, director of the NCI, and Edward Sondik, deputy director, defended their decision in sworn testimony delivered before the Human Resources and Intergovernmental Relations Subcommittee, but not before Representative Towns admonished Broder for withholding information. "Well, let me just say this, you know, Dr. Broder, just so we understand each other," Towns warned. "In my last election I got about 88 percent of the vote. So even if I drop down 10 percent, I still have 78 percent of the votes in my district. So I plan to be around."[61]

For his part, Broder attempted to fend off Towns's accusation that the NCI relied too much on science by reminding his audience that he defined "science" broadly. "I use the term 'science' not in the term of laboratories and laboratory animals," he said. "I use 'science' in the full and best use of that term applying knowledge, orderly rational knowledge to benefit and alleviate science." Broder corrected Towns's assertion that "for 6 years, the National Cancer Institute did not have a valid basis for recommending mammography between the ages of 40 and 49." Noted Broder, "But 80 percent of what was done in 1987 [the year the NCI first issued its under-forty mammography guideline] still stands because approximately 80 percent of the cases of breast cancer are over age 50."[62]

Regarding Towns's argument that the NCI was "rationing women's health" on behalf of insurance companies, Broder responded that there was no such motive at work. "That's not one of the factors that we take into account," he explained bluntly. Despite this testimony, other witnesses continued to hammer Broder's position. Daniel B. Kopans, then director of breast imaging at Massachusetts General Hospital, suggested that a February 1993 NCI-sponsored workshop devoted to reviewing mammogram studies was "loaded." Recounted Kopans, "It's [*sic*] conclusions were predetermined basically what the outcome of the meeting would be." Even Charles R. Smart, retired chief of the NCI's Early Detection Branch, agreed that the "deck was stacked" at the workshop against those who continued to recommend screening of women under fifty.[63]

In its final report on the mammography debate, entitled *Misused Science: The National Cancer Institute's Elimination of Mammography Guidelines for Women in Their Forties*, the House Committee on Government Operations concluded that the NCI "created confusion" and "failed to examine objectively all the scientific evidence on mammography." Further, the committee confirmed allegations that not only had the NCI rigged the revision process to favor elimination of the guideline for women under fifty, but it had "ignored the lack of data on American women, especially minority women." Finally, the report chided the NCI for failing to "maintain records sufficient to allow a public recounting of its decisions and decision-makers."[64]

How did science, even science representing perhaps the best the country could offer, fall into such ill-repute? How could politicians contest scientific expertise in matters of science, and why did scientists disagree so vehemently and publicly with each other? How could opponents agree on the facts and figures yet differ so completely when assessing the risks behind them? How did experts come to relinquish so much authority to laypeople, and why did laypeople view experts with so little esteem? And how did the director of the NCI decide that because "the average woman in this country is very intellectually gifted," she ought to determine for herself whether to get a mammogram before age fifty?[65]

The answer, as anthropologist Mary Douglas points out, was that contemporary culture had politicized the concept of risk, severing its connection to a technical, scientific, and value-free probabilistic construct. Risk without certainty could not operate otherwise. Knowledge, she says, fell apart, replaced by multiple sources of truth. Douglas divided these sources into the "establishment" and "alternative" sources to delineate the older orthodox professional locus of knowledge production from the laissez-faire environment for knowledge production common today, but it clearly became harder and harder to separate the two. Americans, after all, paid more visits to nutrition and herbal supplement dealers and chiropractors than to "mainstream" physicians in 1997. All sources of information effectively became alternative; the "establishment" was driven underground if not already dead. One need only consider the fact that a National Center for Complementary and Alternative Medicine found abode in the National Institutes of Health to see the distinction dissolving.[66]

Alongside alternative knowledge production emerged multiple alternative yardsticks to measure its authenticity. But in a world with no canon, where all knowledge became alternative, where there were no commonly recognized authorities or arbiters, and where there were no intrinsic consequences attached to singular or multiple allegiances, which risk assessment yardstick was best? Wisdom for one became bias for another. The same evidence potentially could produce both "hard science" and "junk science." Complained the authors of a *Forbes* magazine cover story depicting a pig with wings, "In a courtroom, anything will fly if a scientist testifies to it." In a world with no common cultural superstitions, uniform standards of judgment seemed beyond reach.[67]

The problem of alternative yardsticks became glaringly evident in the continuing debate over what constituted a significant or *de minimis* risk. Both the Delaney Cancer Clause and the Clean Air Act established zero-tolerance principles—for artificial chemical additives to food, on the one hand, and for toxic gases, on the other. But as the list of chemicals identified as toxic or carcinogenic grew longer in the 1970s and 1980s, it became a commonplace assumption that life itself must be toxic and carcinogenic. Fully two-thirds of the eight hundred synthetic chemicals that have been subjected to laboratory testing cause tumors to grow in mice. Natural benzene can be found in appreciable quantity in eggs. Twenty-six chemicals in a cup of coffee are known carcinogens in rats, and over a thousand more have yet to be tested.[68]

Against which yardstick can a significant risk be determined? Many scientists ascribed to the "threshold" hypothesis, that chemicals left the body or environment more or less unaffected until they crossed a critical boundary where effects began to show. Some, like the controversial but respected University of California at Berkeley molecular biologist Bruce Ames, argued that "the world is full of poisons [artificial and natural], but it doesn't make any difference" because of the body's natural ability to repair the damage. Others interpreted health risks in logarithmic terms, rising exponentially with each arithmetic increase in amount of offending pollutant. Complicating the matter, the tools and techniques of bioassay became ever more discriminating. Detecting offending chemicals in the parts per billion was considered routine in advanced laboratories and parts per trillion not impossible. Ever-smaller potential risks could now be identified.[69]

The proliferation of alternative yardsticks in the health and biomedical sciences resulted in an even deeper epistemological conundrum. Alternative notions of risk confounded easy comparison of the relative size of risk because there existed no easy interpretation of the term "risk" itself beyond its negative connotation. Different risk managers variously measured risk quantitatively as increased probabilities of death, as reduced life expectancy, and as probability of death per unit of exposure. Each quantitative measure also carried with it one or more qualitative or interpretive components, beginning with "bias" in the choice of measure. In other words, all carried with them significant political baggage.[70]

Alternative yardsticks contributed to public cynicism toward biomedical and health science. Distrust of experts and institutions, particularly government experts and institutions, subsisted easily on a steady supply of contradictory messages flowing from individuals and organizations disseminating alternative information as knowledge. As a consequence, many people, including scientists themselves, began to see science as so much "rhetorical argument" devoid of "fact." Social scientists began uncovering evidence suggesting that the more informed adults were, the more they distrusted science. The effects of widespread distrust were tangible. The number of commissioned public health officers in the Environmental Protection Agency dwindled from over six hundred to only three in the quarter century since 1970, replaced in large measure by attorneys who judge the credibility of risk producers by their behavior, demeanor, and the internal consistency of their arguments rather than on precedent. Sometimes organizations—including corporations—distrusted themselves. The 1999 finding by an independent laboratory hired by Greenpeace of so-called toxic plasticizers in children's chew toys is instructive in this regard. Quoting from the Reuters News Service press release: "Several companies said they believed their products were safe, *but would stop using the chemicals anyway*" (emphasis added).[71]

The highly public and, in this case, nearly real-time caving in of the toy industry is reminiscent of the character of the diethylstilbestrol (DES) debates between cattle ranchers, scientists, feed manufacturers, and drug companies over a hormone supplement that significantly reduced the finishing time and boosted the size of beef cattle. "What was really striking [about the DES debate]," notes historian Alan Marcus, "was the lack of effort on both sides to generate material that might help settle the question." The reaction of the toy makers in light of contemporary notions about knowledge production is understandable. Why bother fighting when all knowledge is alternative?[72]

Contemporary society confronted risk in all this confusion most directly locally and personally. Locally, dueling expertise and escalating stakes dramatically influenced the way many pragmatic Americans attacked the problem of personal risk: They bought health insurance coverage. But insurance itself no longer operated by the principles it did before World War II.

Private health insurers in the early twentieth century allocated service by "community rating," where individuals found coverage under a universal umbrella, truly sharing risk collectively with others like themselves. Though commonly called "social insurance" and often assumed entirely equitable, insured members did not all pay the same amount for health coverage. Individual members initially paid into the system according to a graduated income scale. Insurers covered their expenses by selecting generally healthy groups—sturdy, young salespeople and skilled industrial employees, for example—to insure in the first place. Other groups—like loggers, explosives experts, miners, and construction workers—were ignored and classified as "uninsurable." The pooling or spreading of risk kept rates low enough for those who rarely became sick and high enough to subsidize the costs of those who got sick often. Pooling, when accompanied by careful actuarial work, could and did support retirees who paid in only one dollar for every two paid out in health benefits.[73]

In late-twentieth-century America, however, each individual carried personal risks that increased premiums or led to denial of coverage entirely. Why the change? Consider the traditional axiom that made insurance seem like such a good idea in the first place: "To each according to his needs, from each according to his ability to pay." As Pulitzer Prize–winning historian Paul Starr points out, early health benefit providers like Blue Cross "violated the concept of limited liability and the rule that insurance should never increase the hazard." Strange as it seems in a contemporary world constructed around limits, there were initially no constraints placed on fee bills hospitalized patients might generate. The construction of early health plans, in fact, positively encouraged the insured to seek hospitalization.[74]

However, as health costs escalated (largely because of proliferating expensive but effective treatments) and as individuals became more selfish guardians of their own health, rational arguments emerged against the seemingly outmoded mutual social obligations of community-rated insurance. Health insurers seized this opportunity to build so-called experience rating or merit rating into their premiums. While many Americans continued to secure insurance through their employers, insurers recalibrated and partitioned policies to reflect the "actual loss experience" of individual employees. Lower-risk subscribers like young people or nonsmokers under experience rating received much lower rates than higher-risk groups like the elderly or obese. "Risk groups," defined by insurers, scientists, interest groups, and policy-makers, included those groups at an elevated risk of disease or injury. Risk groups could be formed using racial characteristics, ethnic heritage, family heredity, social behaviors, personal attitudes, and gender as markers. Ironically, experience rating signaled a fragmentation of social order even as it grouped individuals by race, class, and gender.[75]

When measured only in dollars and cents, the benefits of maintaining a healthy body of poor people no longer outweighed the costs shouldered by

Americans with higher incomes and better overall health. These young, relatively prosperous Americans and their employers abandoned older social insurance schemes in rapidly rising numbers for competing commercial providers that targeted low-risk groups by offering significantly lower premiums. Blue Cross belatedly but eagerly followed others in plunging into this new territory, as the alternative meant insuring only the higher-risk groups unwanted by other insurers. As one economist made clear, "In the modern world there is no possibility that those who have the greatest need for the services could ever pay a significant share of the cost from their own resources, even if we considered that fair." The rationale could easily be affirmed in the swelling arguments for personal responsibility in risk-making and risk-taking. Democratic decision-making, as a consequence, no longer guaranteed equal outcomes for all.[76]

HEALTH and longevity in America have never been better than in the past twenty years. The average American can expect to live to the ripe old age of seventy-six, three decades longer than his or her great-grandfather. The dramatic transition in mortality patterns beginning around 1900—from death by infectious disease to death by accumulating chronic diseases—is evidence not of failure in humankind's war against disease but of good fortune. Americans, proportionally speaking, died of cancer, heart disease, and arteriosclerosis because of their striking ability to survive to old age. Newer diseases—Lyme, asbestosis, girdiasis, bovine spongiform encephalopathy, urea formaldehyde poisoning, cancer from Alar, Marburg, and Legionnaires' disease—kill relatively few, if any. They pale alongside the killers of the early republic—smallpox, yellow fever, malaria. Only AIDS threatens to reverse the course, and it, too, is beginning to look a lot like a survivor's disease, as a chronic affliction for those who can afford treatment.[77]

More and a greater percentage of public and private money is devoted to biomedical and health research now than at any other time in history. Between 1950 and 1975 the health research and development budget in this country grew from $160 million to $4.7 billion. By 1989 Americans spent $590 billion in pursuit of better health. Currently, 14 percent of America's gross domestic product is consumed as health care.[78]

Yet the rapid advance in biomedical and health-related expenditures for some time has been seen as part of the problem and not part of the solution. Soaring health care costs prompted Health and Human Services Secretary Joseph Califano to propose that "we are killing ourselves by our own careless habits. . . . Americans can do more for their own health than any doctors, any machine or hospital, by adopting healthy lifestyles." This kind of thinking, cost-containment through inculcation of personal responsibility, has become a permanent fixture of biomedical and health research policy. It was an underlying motivation in the National Health Planning and Resource Development Act of 1974 (PL 93–641). It explains the directive of the *Healthy People 2000* report to create a "culture of character" among people that promotes "responsible behavior" and provides "our best opportunity to reduce the ever-increasing portion of our resources that we spend to treat preventable illness and functional impairment."[79]

At the same time, we have discovered that we can no longer count on medicine for further future radical gains in individual health or longevity. Life expectancy has risen little in recent years and may even be declining for certain groups, like African Americans. The largest gains resulted from declines in infant mortality in the first decades of the twentieth century, and the dividends of that event have now largely been paid. Longevity has its price. Substantial numbers of the elderly are plagued with conditions making life virtually intolerable, opening up the previously unthinkable possibility of assisted suicide as a legally mandated personal choice. The overall precariousness of life remains unaltered.

Moreover, in a world flooded with sometimes contradictory health advice and where health is generally good, taking chances by relying on any particular expert seems foolish. Mistakes and accidents in a society with high expectations are more costly. Today when young people die it is considered a great tragedy, a failure of society to ameliorate suicidal tendencies and devil-may-care attitudes. Less than a century ago, deaths in childhood were commonplace, accepted (if no less painful) human events. Life is rendered more precious even while observation of mortality becomes more infrequent, replaced by virtual experience on television, in films, and in video games.[80]

Risk is itself a cultural and intellectual artifact. It seems unwarranted to question whether biomedical or public health risks in contemporary America are real or imagined, for of course they seem very real to us. A risk perceived is a risk indeed. What we should investigate instead is how the concepts of risk entertained in popular health promotion and biomedical research policy satisfied the requirements of contemporary life. Clearly, contemporary America's notions of risk in relation to the debate over relative personal freedom and diminished public funding were inextricably intertwined with basic and defining cultural and intellectual prerequisites. Americans lived in an age where, justified or not, limits were placed on individual human beings. But at the very same time Americans defined well-being in terms of endless material bounty. No wonder one prominent historian has described late-twentieth-century America as an "Age of Woe."[81]

The newer conception of risk as personal "danger" instead of unavoidable "contingency" neatly reconciled the complicated fantasy of a fragmented contemporary daily life with the complicated collective fantasy of a fragmented contemporary civilization. In contemporary times, expectation bore no relation to, and in fact underwhelmed, the possible. Volumes have been written in recent years about the crying need in this country to restore confidence and credibility in our public institutions and professions, to engage our civic responsibilities once again in order to satisfactorily confront the inequalities of health, but it seems evident that we have gotten along quite well by subverting them to contemporary cultural priorities. Risk assessment, in sum, revealed much about our alienated selves and our brooding, apprehensive society and perhaps less about the overall safety and health of that society.[82]

Notes

The author wishes to thank Harvey M. Sapolsky and Alan I Marcus for their comments and kind help in the preparation of this chapter.

1. Howard M. Leichter, *Free to Be Foolish: Politics and Health Promotion in the United States and Great Britain* (Princeton, N.J.: Princeton University Press, 1991), 9–10, 13; Mary Douglas, "Risk as a Forensic Resource: From 'Chance' to 'Danger,'" *Daedalus* 119 (fall 1990): 1–3; Baruch Fischhoff, "Managing Risk Perceptions," *Issues in Science and Technology* (fall 1985): 89; Walter B. Wriston, "Risk and Other Four-Letter Words," *Vital Speeches of the Day* 46 (December 15, 1979): 158. See also Anthony Simones, "The Right to Suffer as Individuals or the Necessity to Survive as a Society: The Right of Privacy and AIDS" (paper presented at the meeting of the Western Social Science Association, Ft. Worth, Tex., 1999); William Graebner, *The Age of Doubt: American Thought and Culture in the 1940s* (Boston: Twayne, 1991).
2. U.S. Congress, Office of Technology Assessment, *Researching Health Risks* (Washington, D.C.: Government Printing Office, 1993), 5–7, 11; Gilbert S. Omenn, "Making Use of Cancer Risk Assessment," *Issues in Science and Technology* (summer 1996): 29–32; Department of Health and Human Services, Task Force on Health Risk Assessment, *Determining Risks to Health* (Dover, Mass.: Auburn House, 1986), 69, 123, 172.
3. National Research Council, Commission on Life Sciences, Committee on the Institutional Means for Assessment of Risks to Public Health, *Risk Assessment in the Federal Government: Managing the Process* (Washington, D.C.: National Academy Press, 1983), ix, 5; Gail Charnley, "My Risk Is Greater Than Your Risk," *Vital Speeches of the Day* 64 (July 15, 1998): 585.
4. National Institute of Medicine, Division of Health Care Services, Division of Health Sciences Policy, Committee on Xenograft Transplantation, *Xenotransplantation: Science, Ethics, and Public Policy* (Washington, D.C.: National Academy Press, 1996).
5. Ibid., 10–16, 39–41, 92–96.
6. Ibid., 2–5.
7. Elizabeth Pennisi, "FDA Panel OKs Baboon Marrow Transplant," *Science* 269 (July 21, 1995): 293–294; William H. Allen, "AIDS Treatment to Use Baboon Cells," *BioScience* 45 (June 1995): 391.

8. Lawrence K. Altman, "Man Gets Baboon Marrow in Risky AIDS Treatment," *New York Times*, December 15, 1995; Lawrence K. Altman, "So Far, So Good for Baboon Marrow Patient," *New York Times*, December 16, 1995; Pennisi, "FDA Panel OKs Baboon Marrow Transplant," 293; Christine Gorman, "Taking a Big Risk for a Cure," *Time*, January 15, 1996, 59.
9. Jonathan S. Allan, "Should Transplants from Monkeys to Humans Be Permitted?" *Health* (October 1995): 22; Jonathan S. Allan, "Fear of Viruses," *New York Times*, January 20, 1996; Lawrence K. Altman, "When Doctors and Patients Decide to Test the Far Limits of Treatment," *New York Times*, December 19, 1995; Pennisi, "FDA Panel OKs Baboon Marrow Transplant," 293.
10. Gina Kolata, "Transplant: Urgent Step or Step Off the Edge?" *New York Times*, January 9, 1996.
11. Ibid.; Rachel Nowak, "FDA Puts the Brakes on Xenotransplants," *Science* 268 (May 5, 1995): 630.
12. Lawrence K. Altman, "Baboon Cells Fail to Thrive, but AIDS Patient Improves," *New York Times*, February 9, 1996.
13. Mary Douglas and Aaron Wildavsky, *Risk and Culture: An Essay on the Selection of Technical and Environmental Dangers* (Berkeley: University of California Press, 1982), 21.
14. Ibid., 48; Richard L. Meehan, *The Atom and the Fault: Experts, Earthquakes, and Nuclear Power* (Cambridge: MIT Press, 1984), 138.
15. Douglas and Wildavsky, *Risk and Culture*, 19, 69–71; Alan I Marcus, *Cancer from Beef: DES, Federal Food Regulation, and Consumer Confidence* (Baltimore: Johns Hopkins University Press, 1994), 7, 159.
16. Alvin M. Weinberg, "Science and Its Limits: The Regulator's Dilemma," *Issues in Science and Technology* (fall 1985): 59–71; John N. Lavis and Gregory L. Stoddart, "Can We Have Too Much Health Care?" *Daedalus* 123 (fall 1994): 49.
17. Paul Starr, "Medicine and the Waning of Professional Sovereignty," *Daedalus* 107 (winter 1978): 175; Theodore J. Lowi, "Risks and Rights in the History of the American Governments," *Daedalus* 119 (fall 1990): 19, 38–39.
18. John H. Knowles, "The Responsibility of the Individual," *Daedalus* 106 (winter 1977): 58–59; Lynn E. Ponton, "Their Dark Romance with Risk," *Newsweek*, May 10, 1999, 55.
19. John H. Knowles, "Introduction," *Daedalus* 106 (winter 1977): 6; Daniel Callahan, "Health and Society: Some Ethical Imperatives," *Daedalus* 106 (winter 1977): 32. See also National Institute of Medicine, Division of Mental Health and Behavioral Medicine, *Behavior, Health Risks, and Social Disadvantage* (Washington, D.C.: National Academy Press, 1982).
20. Robert K. Oldham, "Patients and Research Partners," *Vital Speeches of the Day* 53 (October 1, 1987): 764–765; Claudia Kalb, "To Be a Guinea Pig," *Newsweek*, October 19, 1998, 86.
21. Knowles, "Responsibility of the Individual," 60, 62, 78; Ellyn E. Spragins, "So What's the Score?" *Newsweek*, October 19, 1998, 90.
22. Marie R. Haug and Bebe Lavin, "Practitioner or Patient: Who's in Charge?" *Journal of Health and Social Behavior* 22 (September 1981): 212–213; Renée C. Fox, "The Medicalization and Demedicalization of American Society," *Daedalus* 106 (winter 1977): 19.
23. Howard R. Veit, "Health Care Consumerism: Its Impact on Health Care Providers in the 21st Century," *Vital Speeches of the Day* 64 (July 1, 1998): 565–566.
24. Allan M. Brandt, *"Just Say No": Risk, Behavior, and Disease in Twentieth-Century America* (Cambridge: MIT, Cultural Studies Project), 13–15, 21–22; Robert G.

Evans, "Health Care as a Threat to Health: Defense, Opulence, and the Social Environment," *Daedalus* 123 (fall 1994): 21–42.

25. Starr, "Medicine and the Waning of Professional Sovereignty," 175–193; John C. Burnham, "American Medicine's Golden Age: What Happened to It?" *Science* 215 (March 19, 1982): 1474–1479.
26. Harvey M. Sapolsky, introduction to *Consuming Fears: The Politics of Product Risks*, ed. Harvey M. Sapolsky (New York: Basic Books, 1986), 4; George Gerbner, "Science on Television: How It Affects Public Conceptions," *Issues in Science and Technology* (spring 1987): 109–115; M. J. Sandow, "Suprascapular Nerve Rotator Cuff Compression Syndrome in Volleyball Players," *Journal of Shoulder and Elbow Surgery* 7 (September–October 1998): 516–521; Lori Lichay, *Predictive Validity of the Lateral Scapular Slide Test for Early Season Injuries in Overhead Athletes* (Springfield, Mass.: N.p., 1996); Steven R. Tippett, "Objectivity and Validity of the Lateral Scapular Slide Test" (M.S. thesis, Illinois State University, 1995); Rebecca Lynn Stine, "The Effect of Eccentrically Training the Scapular Stabilizer Muscles on the Ball Velocity During a Tennis Serve" (M.S. thesis, University of Kentucky, 1992); Sarah Elizabeth Alexander, "Isokinetic Strength Measures of Shoulder Rotation in the Plane of the Scapula: A Descriptive Study of Healthy Individuals" (M.S. thesis, North George College and State University, 1996); Wendy W. Weil, "Strength Training to Prevent 'Swimmer's Shoulder,'" *Swimming World* 40 (August 1, 1999): 19; Wendy W. Weil, "Preventing Swimmer's Shoulder," *Swimming World* 40 (July 1, 1999): 29; Francisco Grande, "Assessment of Body Fat in Man," in *Obesity in Perspective*, ed. George A. Bray (Washington, D.C.: Government Printing Office, 1976), 189–203.
27. On the history of obesity, see Laura Fraser, *Losing It: America's Obsession with Weight and the Industry That Feeds on It* (New York: Dutton, 1997); Hillel Schwartz, *Never Satisfied: A Cultural History of Diets, Fantasies, and Fat* (New York: Free Press, 1986); Peter Stearns, *Fat History: Bodies and Beauty in the Modern West* (New York: New York University Press, 1997).
28. Ancel Keys, "Obesity Measurement and the Composition of the Body," in *Overeating, Overweight, and Obesity: Proceedings of the Nutrition Symposium Held at the Harvard School of Public Health, Boston, Massachusetts, October 29, 1952* (New York: National Vitamin Foundation, 1953), 13–14; Rachel Schemmel, "Assessment of Obesity," in *Nutrition, Physiology, and Obesity*, ed. Rachel Schemmel (Boca Raton, Fla.: CRC Press, 1980), 3; Donald B. Armstrong, "What Is Normal Weight for Health?" in *Your Weight and How to Control It*, ed. Morris Fishbein (Garden City, N.Y.: Doubleday, 1949), 1–17.
29. David S. Gray et al., "Skinfold Thickness Measurements in Obese Subjects," *American Journal of Clinical Nutrition* 51 (April 1990): 571–577; Aviva Must et al., "Reference Data for Obesity: 85th and 95th Percentiles of Body Mass Index and Triceps Skinfold Thickness," *American Journal of Clinical Nutrition* 53 (April 1991): 839–846; see also J. P. Clarys et al., "The Skinfold: Myth and Reality," *Journal of Sports Sciences* 5 (1987): 3–33.
30. E. E. Hunt, "Factors in Human Obesity," in *Overeating, Overweight, and Obesity*, 73–89.
31. Paul Ernsberger and P. Haskew, "Rethinking Obesity," *Journal of Obesity and Weight Regulation* 6 (1987): 58–137.
32. George Christakis, "The Prevalence of Obesity," in Bray, *Obesity in Perspective*, 209. See especially Laura S. Brown and Ether D. Rothblum, *Overcoming Fear of Fat* (New York: Harrington Park Press, 1989); Roberta Pollack Seid, *Never Too Thin: Why Women Are at War with Their Bodies* (New York:

Prentice-Hall, 1989); Ruth Raymond Thone, *Fat: A Fate Worse Than Death? Women, Weight, and Appearance* (New York: Haworth Press, 1997); Jaclyn Packer, "The Role of Stigmatization in Fat People's Avoidance of Physical Exercise," *Women and Therapy* 8 (March 1989): 49–63; Angela Barron McBride, "Fat Is Generous, Nurturing, and Warm . . . ," *Women and Therapy* 8 (March 1989): 93–103. See also Marvin Grosswirth, *Fat Pride: A Survival Handbook* (New York: Jarrow Press, 1971).

33. George A. Bray, "Overweight Is Risking Fate: Definition, Classification, Prevalence, and Risks," *Annals of the New York Academy of Sciences* 499 (1987): 14.
34. B. T. Burton et al., "Health Implications of Obesity: An NIH Consensus Development Conference," *International Journal of Obesity* 9 (1985): 155–170; "Health Implications of Obesity: National Institutes of Health Consensus Development Conference Statement," *Annals of Internal Medicine* 103 (1985): 147–151; Eleanor D. Schlenker, "Obesity and the Lifespan," in Schemmel, *Nutrition, Physiology, and Obesity*, 152–166; Jules Hirsch, "The Psychological Consequences of Obesity," in Bray, *Obesity in Perspective*, 81–102.
35. Department of Health and Human Services, Public Health Service, *Healthy People 2000: National Health Promotion and Disease Prevention Objectives* (Washington, D.C.: Government Printing Office, 1991), 96, 114–116, 403, 461; Donald Budd Armstrong, "Obesity and Its Relation to Health and Disease," *Journal of the American Medical Association* 147 (November 10, 1951): 1007–1014; Donald Budd Armstrong et al., "Influence of Overweight on Health and Disease," *Postgraduate Medicine* 10 (November 1951): 407–421; "Ideal Weights for Women," *Statistical Bulletin, Metropolitan Life Insurance Company* 23 (October 1942): 6–8; "Ideal Weights for Men," *Statistical Bulletin, Metropolitan Life Insurance Company* 24 (June 1943): 6–8; "What's Your BMI?" *CQ Researcher* (January 15, 1999): 32; William Stinner et al., "Redefining Federal Obesity Guidelines: Subgroup Variations in Obesity Likelihood in Utah Employing Older and New Definitions" (paper presented at the meeting of the Western Social Science Association, Ft. Worth, Tex., 1999); John Foreyt and Ken Goodrick, "The Ultimate Triumph of Obesity," *Lancet* 346 (1995): 134–135.
36. Sharon Dalton, introduction to *Overweight and Weight Management: The Health Professional's Guide to Understanding and Practice* (Gaithersburg, Md.: Aspen, 1997), xix; B. G. Charlton and M. J. Tovee, "Nomenclature of Optimal BMI: Slim's the Word," *Quarterly Journal of Medicine* 92 (July 1, 1999): 418; M. Doyle, "Can Patients' Knowledge of Their Own Weight and Height Be Used to Replace Measured Height and Weight in the Calculation of BMI?" *Proceedings of the Nutrition Society* 57 (1998): 165A.
37. American Cancer Society, *Cancer Facts and Figures, 1991* (Atlanta: American Cancer Society, 1991); American Cancer Society, *Cancer Facts and Figures, 1997–1998* (Atlanta: American Cancer Society, 1998).
38. American Cancer Society, *Cancer Facts and Figures, 1991*, 52; U.S. Congress, Senate, Committee on Labor and Human Resources, Subcommittee on Aging, *Improving the Quality of Mammography: How Current Practice Fails* (102nd Cong.), February 13, 1991 (S. Hrg. 1193), 23. See also U.S. Congress, House, Committee on Commerce, Subcommittee on Health and Environment, *Reauthorization of the Mammography Quality Standards Act* (105th Cong., 2nd sess.), May 8, 1998 (Serial no. 88).
39. Phyllida Brown, "Breast Cancer Drug Hits Fresh Safety Snag," *New Scientist*, March 14, 1992, 12; Janet Raloff, "Tamoxifen Quandary: Promising Drug May Hide a Troubling Dark Side," *Science News*, April 25, 1992, 266–269; U.S. Congress, House, Committee on Government Operations, Human Resources and Intergov-

ernmental Relations Subcommittee, *Breast Cancer Prevention Study: Are Healthy Women Put at Risk by Federally Funded Research?* (102nd Cong., 2nd sess.), October 22, 1992, 1–2, 93.

40. U.S. Congress, *Breast Cancer Prevention Study*, 6. See also Tommy Fornander et al., "Adjuvant Tamoxifen in Early Breast Cancer: Occurrence of New Primary Cancers," *Lancet* 1 (January 21, 1989): 117–120; Adriane Fugh-Berman and Samuel Epstein, "Tamoxifen: Disease Prevention or Disease Substitution?" *Lancet* 340 (November 7, 1992): 1143–1145; Christopher J. Jolles, "Cystic Ovarian Necrosis Complicating Tamoxifen Therapy for Breast Cancer in a Premenopausal Woman: A Case Report," *Journal of Reproductive Medicine* 35 (March 1990): 299–300; Nicholas A. Pavlidis et al., "Clear Evidence That Long-Term, Low-Dose Tamoxifen Treatment Can Induce Ocular Toxicity," *Cancer* 69 (June 15, 1992): 2961–2964.
41. Nancy Bruning, "Tamoxifen on Trial: Damned If We Do, Damned If We Don't," *Breast Cancer Action* 13 (August 1992): 1; U.S. Congress, *Breast Cancer Prevention Study*, 2–3, 6, 46.
42. U.S. Congress, *Breast Cancer Prevention Study*, 27, 39, 54.
43. Ibid., 19, 63–65.
44. Ibid., 69–71.
45. Ibid., 74–86, 100. Compare Peck's statement with that of Arthur Caplan reported by Janet Raloff in the pages of *Science News*, November 28, 1992, 378: "Most people 'neither want nor expect to live in a risk-free world,' Caplan observes. 'Americans are strongly committed to the view that each person must decide what sorts of risks and hazards they want to face in the service of attaining goals they hold dear.'"
46. U.S. Congress, *Breast Cancer Prevention Study*, 62, 138–152.
47. "National Institutes of Health/National Cancer Institute Consensus Development Meeting on Breast Cancer Screening: Issues and Recommendations," *Journal of the National Cancer Institute* 60 (1978): 1519–1521; "NIH/NCI Consensus Development Meeting on Breast Cancer Screening: Issues and Recommendations," *Yale Journal of Biology and Medicine* 51 (1978): 3–7; Gina Kolata, "Mammogram Talks Prove Indefinite," *New York Times*, January 24, 1997; U.S. Congress, House, Committee on Government Operations, Human Resources and Intergovernmental Relations Subcommittee, *National Cancer Institute's Revision of Its Mammography Guidelines* (103rd Cong., 2nd sess.), March 8, 1994, 2.
48. Gina Kolata, "Mammogram Guideline Is Dropped," *New York Times*, December 5, 1993.
49. Ibid.
50. Ibid.
51. Ibid.; Daniel B. Kopans, "An Overview of the Breast Cancer Screening Controversy," *Journal of the National Cancer Institute Monographs* 22 (1997): 2; U.S. Congress, *National Cancer Institute's Revision of Its Mammography Guidelines*, 140–141.
52. Kolata, "Mammogram Guideline Is Dropped."
53. Joyce Purnick, "The Mammogram Controversy: It Confuses Women, Raises Unfounded Fears," *New York Times*, December 27, 1993.
54. Several months later, in his testimony before the Human Resources and Intergovernmental Relations Subcommittee on Government Operations of the U.S. House of Representatives, NCI director Sam Broder defended the decision to advise against mammograms in women under fifty in part because "it does expose them to ionizing radiation." See U.S. Congress, *National Cancer Institute's Revision of Its Mammography Guidelines*, 103.
55. U.S. Congress, *National Cancer Institute's Revision of Its Mammography Guidelines*, 69–70.

56. Ibid., 2.
57. Ibid.
58. Ibid., 3–4.
59. Ibid., 4–5, 32–33, 35.
60. Ibid., 14–15, 19, 21–22, 33.
61. Ibid., 71.
62. Ibid., 75, 102–103.
63. Ibid., 110–111, 172–174.
64. U.S. Congress, House, Committee on Government Operations, *Misused Science: The National Cancer Institute's Elimination of Mammography Guidelines for Women in Their Forties* (103rd Cong., 2nd sess.), October 24, 1994 (H. Rept. 863), 7–17.
65. *National Cancer Institute's Revision of Its Mammography Guidelines*, 111–112, 120.
66. Douglas and Wildavsky, *Risk and Culture*, 8–9, 73, 186; Douglas, "Risk as a Forensic Resource," 11–12; Geoffrey Cowley and Anne Underwood, "What's 'Alternative'?" *Newsweek*, November 23, 1998, 68; National Institutes of Health, *Alternative Medicine: Expanding Medical Horizons: A Report to the National Institutes of Health on Alternative Medical Systems and Practices in the United States, Prepared Under the Auspices of the Workshop on Alternative Medicine, Chantilly, Virginia, September 14–16, 1992* (Washington, D.C.: Government Printing Office, 1995). See also James Harvey Young, "The Development of the Office of Alternative Medicine in the National Institutes of Health, 1991–1996," *Bulletin of the History of Medicine* 72 (summer 1998): 279–298.
67. Lavis and Stoddart, "Can We Have Too Much Health Care?" 46.
68. Daniel Byrd and Lester B. Lave, "Narrowing the Range: A Framework for Risk Regulators," *Issues in Science and Technology* (summer 1987): 92; James Trefil, "How the Body Defends Itself from the Risky Business of Living," *Smithsonian* (December 1995): 43–49.
69. John F. Ross, "Risk: Where Do Real Dangers Lie?" *Smithsonian* (November 1995): 48; Trefil, "How the Body Defends Itself," 43–49; Alvin M. Weinberg, "Science and Its Limits: The Regulator's Dilemma," *Issues in Science and Technology* (fall 1985): 66; Elizabeth M. Whelan, "Health Hoax and a Health Scare: Why Americans Don't Know the Difference," *Vital Speeches of the Day* 54 (November 1, 1987): 59.
70. Harvey M. Sapolsky, "The Politics of Risk," *Daedalus* 119 (fall 1990): 83; Harvey M. Sapolsky, "The Politics of Product Controversies," in *Consuming Fears: The Politics of Product Risks*, ed. Harvey M. Sapolsky (New York: Basic Books, 1986), 187; Douglas and Wildavsky, *Risk and Culture*, 1, 4; Douglas, "Risk as a Forensic Resource," 12; David J. Rothman, *Strangers at the Bedside: A History of How Law and Bioethics Transformed Medical Decision Making* (New York: Basic Books, 1991), 10.
71. Honey Rand, "Science, Non-Science, and Nonsense," *Vital Speeches of the Day* 64 (February 15, 1998): 282–284; Weinberg, "Science and Its Limits," 68; J. Kluger, "Poisonous Plastics?" *Time*, March 1, 1999, 53; M. L. Marin et al., "Analysis of Potentially Toxic Phthalate Plasticizers Used in Toy Manufacturing," *Bulletin of Environmental Contamination and Toxicology* 60 (January 1998): 68–73.
72. Marcus, *Cancer from Beef*, 6.
73. Rashi Fein, *Medical Care, Medical Costs: The Search for a Health Insurance Policy* (Cambridge: Harvard University Press, 1986), 28–33, 53, 219.
74. Paul Starr, *The Social Transformation of American Medicine* (New York: Basic Books, 1982), 298.
75. Duncan M. MacIntyre, *Voluntary Health Insurance and Rate Making* (Ithaca, N.Y.: Cornell University Press, 1962), 26–58; Herman Miles Somers and Anne Ramsay

Somers, *Doctors, Patients, and Health Insurance: The Organization and Financing of Medical Care* (Garden City, N.Y.: Anchor Books, 1962), 309–327. See also Louise B. Russell, *Is Prevention Better Than Cure?* (Washington, D.C.: Brookings Institution, 1986).

76. Robert G. Evans, "Health Care as a Threat to Health: Defense, Opulence, and the Social Environment," *Daedalus* 123 (fall 1994): 24.
77. Alan I Marcus and Hamilton Cravens, introduction to *Health Care Policy in Contemporary America*, ed. Alan I Marcus and Hamilton Cravens (University Park: Pennsylvania State University Press, 1997), 1–4; Lewis Thomas, "On the Science and Technology of Medicine," *Daedalus* 106 (winter 1977): 35–36.
78. Knowles, "Introduction," 3; Office of Technology Assessment, *Health Care Reform* (Washington, D.C.: Government Printing Office, n.d.); Leichter, *Free to Be Foolish*, 5.
79. Theodore R. Marmor and Richard W. Smithey, "Health Policy in Historical Perspective: A Review Essay," *Journal of Policy History* 1 (1989): 114; Department of Health and Human Services, *Healthy People: The Surgeon General's Report on Health Promotion and Disease Prevention* (Washington, D.C.: Government Printing Office, 1979); Drew Altman et al. *Health Planning and Regulation: The Decision-Making Process* (Washington, D.C.: AUPHA Press, 1981), ix; Department of Health and Human Services, *Healthy People 2000*, v, 1–3.
80. Kathy J. Helzlsouer and Leon Gordis, "Risks to Health in the United States," *Daedalus* 119 (fall 1990): 193, 202; Douglas and Wildavsky, *Risk and Culture*, 11; Sapolsky, "The Politics of Product Controversies," 199.
81. Nils-Eric Sahlin and Johannes Persson, "Epistemic Risk: The Significance of Knowing What One Does Not Know," in *Future Risks and Risk Management*, ed. Berndt Brehmer and Nils-Eric Sahlin (Boston: Kluwer Academic Publishers, 1994), 37–62; Paul Slovic, "Perceptions of Risk: Paradox and Challenge," in Brehmer and Sahlin, *Future Risks*, 63–78; Thomas, "On the Science and Technology of Medicine," 35–46.
82. Marcus and Cravens, introduction to *Health Care Policy*, 4; Douglas and Wildavsky, *Risk and Culture*, 14; Stephen R. Graubard, "Preface," *Daedalus* 119 (fall 1990): v. See also Mary Douglas, *Risk Acceptability According to the Social Science: Occasional Reports on Current Topics* (New York: Russell Sage Foundation, 1985).

Chapter 5

Progress and Its Discontents

Postwar Science and Technology Policy

HOWARD P. SEGAL

IN THE INTRODUCTION to his seminal 1945 report, *Science—The Endless Frontier*, Vannevar Bush wrote that "we have no national policy for science. The Government has only begun to utilize science in the Nation's welfare. There is no body within the Government charged with formulating or executing a national science policy. . . . Science has been in the wings. It should be brought to the center of the stage—for in it lies much of our hope for the future."[1] Five years after the report appeared, and following extensive debate inside and outside of Congress and the White House, the National Science Foundation (NSF) finally came into being.

In its early formulations and applications, postwar science and technology policy largely excluded the social sciences. Bush himself had no use for the social sciences, which he deemed mushy, propagandistic, and, most damning, unscientific. Bush's influence on immediate postwar science and technology policy overall was so great that his characterization of the social sciences pervaded many quarters. To be sure, Bush was hardly alone in that assessment, and one still hears similar comments from some "pure" scientists.

But Bush likewise excluded "applied science" from the intended focus of postwar federal support, which would instead go to "basic science," or biology, chemistry, and physics. By "applied science" Bush meant engineering. This was not because Bush was either uninterested in the application of scientific discoveries or insensitive to America's historic "practical culture."[2] As a distinguished MIT electrical engineering Ph.D. recipient, professor, department chair, and eventually vice president and dean of engineering, Bush had more than a passing interest in both. Indeed, he himself had many inventions and patents to his credit and became the leading designer of electromechanical analog computers.

Yet Bush's own career and his prewar and wartime experiences had impressed on him the value—but also the unpredictable path—of basic science for ultimately practical purposes. Prominent wartime examples included basic research in nuclear physics leading to the atomic bomb and in microbiology leading to penicillin. Better to let private industry operating in the free marketplace fund avowedly applied science and to save precious federal funds for the kind of basic science that private industry ordinarily wouldn't fund.[3] As Bush put it repeatedly, "applied research invariably drives out pure."[4] Hence the name National *Science* Foundation, not National Science *and Engineering* Foundation (though Bush actually wanted the name National *Research* Foundation).

Bush thereby perpetuated in a monumental way the modern hierarchy that places science above technology; that treats engineers and other technical experts as handmaidens to scientists; that deems technology merely applied science; and that thereby misses the unique intellectual qualities of technology itself. As Princeton University civil engineer David Billington has argued, "Science is discovery, engineering is design. Scientists study the natural, engineers create the artificial. Scientists create general theories out of observed data; engineers make things, often using only very approximate theories." Engineers' "primary motive for design is the creation of an object that works."[5] Or as Stanford aeronautical engineer Walter Vincenti has demonstrated in case studies from aeronautical history, whereas the knowledge generated by scientists is ordinarily used by them to generate more knowledge, the knowledge generated by engineers is ordinarily used by them to design artifacts; if engineering knowledge is used to generate more knowledge, that is a quite secondary objective.[6] Hence the false foundation of that modern hierarchy that Bush embraced. If, however, "applied science" did not in Bush's and other powerbrokers' views warrant anywhere as much federal support as "pure science," what claims on federal largesse could the social sciences possibly make? And how could the social sciences ever justify their own intellectual stature?

By contrast, Bush's principal antagonist, Senator Harley Kilgore of West Virginia, pushed for large federal spending in the social sciences as well as in basic (and applied) science. A New Deal liberal, Kilgore linked the social sciences with direct assistance to ordinary citizens, as, for example, with better housing and better public schools. To him, the benefits of expanded science and technology research would likewise filter down to the grass roots by the same route. But the final legislation favored Bush, not Kilgore, save, ironically, in the naming of the new organization, Kilgore having proposed National Science Foundation.[7]

By the late 1950s and early 1960s, however, things had turned around for the social sciences and so in turn for their use in science and technology policy. To begin with, the self-confidence by then enjoyed by American science and, to a growing degree, American technology in a time of unprecedented federal support and citizen enthusiasm allowed increasing room on the policy pedestal for the social sciences. Indeed, some prominent social scientists themselves virtually shaped debates over public policy and so justified their own and fellow social scientists' increasing respect and power. For example, economist John Kenneth Galbraith's best-selling *The Affluent Society* (1958) convinced many not only that affluence had by now pervaded all of American society, save for pockets of poverty, but also that major economic and social problems either had been or soon would be solved. Moreover, professional economists in policy-making roles were, according to Galbraith and others, heavily responsible for this persistent prosperity. Even the belated discovery of persistent poverty in Appalachia and other areas dominated by whites—as detailed in Michael Harrington's *The Other America* (1962)—followed by President Lyndon Johnson's War on Poverty, did not diminish this optimism that economists could attain affluence for all. If anything, the Council of Economic Advisers under both President John F. Kennedy and President Johnson believed in their ability to develop policies to help to eradicate poverty.

Similarly, sociologist Daniel Bell's influential *The End of Ideology* (1960), however misread as an uncritical endorsement of this consensus over the alleged end of strongly ideological politics in affluent America, nevertheless gave enormous intellectual legitimacy to it. Furthermore, that purported consensus on America's present and future was now traced to the nation's past by influential historians like Daniel Boorstin in *The Genius of American Politics* (1953) and David Potter in *People of Plenty: Economic Abundance and the American Character* (1954). Both discovered an American history allegedly marked by far more agreement on fundamental values and policies than disagreement, and both deemed dissenters from that consensus marginal figures if not outright subversives. Neither Boorstin nor Potter claimed to be a social scientist, but Potter in particular based much of his analysis of the "American Character" on the findings of social scientists.

In this same period the social sciences also played a crucial role in foreign policy, one more directly related to science and technology. Political scientists, sociologists, economists, and others provided expertise and, not least, models for the maturation of "underdeveloped" countries in Latin America, Africa, and Asia into more industrialized and more democratized nations emulating the United States. At the peak of the Cold War, such models were invaluable political as much as academic tools. Area studies, international studies, and development studies became prominent interdisciplinary subfields. Overarching them was "modernization" theory. "Modernization" became the buzzword of politicians and social scientists alike, and both groups believed that the exportation to these backward lands of American science and technology—agricultural chemicals and equipment, for instance—would turn the tide within them from Communism to capitalism or undermine any efforts to remain ideologically neutral. Using a very traditional notion of "technological determinism," and ignorant of historical contexts that might have tempered their optimism, these politicians and social scientists assumed that the sheer transfer abroad of American science and technology would steadily alter centuries-old native cultures in the direction of American culture, not least American democracy. The United States's own revolutionary past was largely overlooked amid the expectation that such profound political as well as cultural change could somehow be managed peacefully. For that matter, repeated examples of American intervention in the internal affairs of other countries from the late nineteenth century through the Cold War were conveniently ignored, as were persistent American leanings toward business, military, and political forces in those countries often hostile toward genuine grass-roots democracy. Moreover, these remarkably shallow and frankly unscientific premises were papered over with elaborate, self-consciously "scientific" models—models replete with graphs, statistics, public opinion surveys, and other seemingly objective paraphernalia. Like their conventional scientific counterparts, these social scientific experiments were designed to be repeated. But the laboratories for "modernization" theory were entire countries. More often than not, however, the experiments failed.[8]

These notable contributions by social scientists to public policy both at home and abroad, combined with the apparent national consensus over values and vision, made it far less risky than in prior eras to employ the social sciences in domestic science and technology policy. Even the most skeptical scientists and engineers had to concede this. Fairly or not, the social sciences had once been associated with the Great Depression as part of the failed and polarizing policies of President Herbert Hoover—the first president to use them in the formulation of public policy, as in *Recent Social Trends in the United States* (1933).[9] The book's publication after Hoover's decisive loss to Franklin Roosevelt in his reelection bid naturally minimized not only its influence but also its pioneering analyses of topics ranging from communications to crime, from education to religion, and from recreation to taxation. By contrast, by the 1950s the social sciences were associated with ongoing national prosperity and international prestige leading in turn to ever-greater federal funding of science and technology. In fact, in the form of public opinion surveys of unprecedented accuracy, the social sciences began to make a case for their own importance by proving how supportive Americans generally were about that greater funding and how much of a consensus there really was about values and vision.

In a deeper sense, and in an ironic reversal of roles, the social sciences now provided intellectual legitimacy for the scientists, engineers, politicians, and others who needed to convince the general public that American science and technology policy would benefit all citizens in more and more direct ways than Vannevar Bush had contended. By the early 1960s it had become an article of faith among liberal politicians, bureaucrats, and social scientists—that is, those eager to use the powers of government to effect economic and social change—that knowledge could and should be deployed to ameliorate human life and to solve those remaining problems of American and, in due course, every other "modernized" society. To be sure, this belief dated to the Enlightenment (the so-called Enlightenment Project) and was hardly restricted to the United States or to other Western capitalist countries, as exemplified by its simultaneous popularity in the Soviet Union—or by its prior popularity in Nazi Germany, as detailed in Jeffrey Herf's *Reactionary Modernism* (1984). The persuasiveness of the Enlightenment Project was not altogether surprising, however, insofar as there was an unacknowledged convergence of belief among many capitalists, communists, and fascists that the supremacy of the powers of rational knowledge and technical skills to improve life rested precisely on their alleged universality and efficiency alike. As some Americans recalled, with considerable satisfaction, even Lenin had tried to adopt Henry Ford's assembly line techniques and Frederick Winslow Taylor's scientific management schemes back in the 1920s.[10]

But nowhere was this belief in the power of knowledge practiced more persistently than in the Kennedy and Johnson administrations and in the colleges and universities where its adherents taught and wrote. Those presidents' "best and brightest" advisers, like Robert McNamara and McGeorge Bundy, are obvious examples of such true believers, but more interesting cases, in some ways, are political scientists Robert Dahl and Robert Lane and nuclear engineer Alvin Weinberg.

In *Who Governs? Democracy and Power in an American City* (1961) and other works, Dahl argued that pluralism was an American fact of life as much as an American political theory and that every interest group that wished to participate in politics at every level had relatively equal opportunity to do so; given the nation's growing affluence, there was no need for alienation from politics but likewise no need for political upheaval. As Dahl put it in *A Preface to Democratic Theory* (1956), "A central guiding thread of American constitutional development has been the evolution of a political system in which all the active and legitimate groups in the population can make themselves heard at some crucial stage in the process of decision."[11]

Lane agreed with Dahl but went further. In such writings as "The Politics of Consensus in an Age of Affluence" (1965) and "The Decline of Politics and Ideology in a Knowledgeable Society" (1966), he contended that traditional ideological politics would decline as that increasing national affluence made the outcomes of political struggles ever less significant and, no less important, as neutral information, rational analysis, and technical skills took their place. If, Lane argued, "one thinks of a domain of 'pure politics' where decisions are determined by calculations of influence, power, or electoral advantage," how much better will be the emerging "domain of 'pure knowledge' where decisions are determined by calculations of how to implement agreed-upon values with rationality and efficiency."

Lane conceded that "there will always be politics; there will always be rationalized self-interest, mobilized by interest groups and articulated in political parties." But he fully expected that such crass "political criteria" for determining public policy would steadily give way to "more universalistic *scientific* criteria." Not laypeople, much less conventional politicians, but rather "professional problem-oriented *scientists*"—and, in effect, social scientists like Lane—would now hold sway.[12] Indeed, Lane offered examples of "scientific" findings about American society by social scientists that led to the kind of public policy activism by experts that he favored: curbing infant mortality, reducing cancer in children, raising individuals and families above the subsistence level, and increasing respect for different racial, religious, and ethnic groups. Such scientific "knowledge—discovered, organized, and communicated by professionals—creates a pressure for policy change with a force all its own."[13]

Weinberg went further still. In "Can Technology Replace Social Engineering?" (1966), he offered examples of how technology could readily find shortcuts—or "Quick Technological Fixes"—to solve social problems that ideological politics could not or would not solve. His examples included free air conditioners and free electricity to reduce the discomfort that in part led to summertime riots in urban ghettos; cheap computers to replace rather than retrain inadequate teachers in impoverished elementary schools; intrauterine devices to reduce population in overpopulated countries where large families are the norm; nuclear-powered desalination to provide freshwater in needy areas where water conservation is difficult if not impossible; safer cars to reduce accidents caused by bad drivers who would resist formal driver instruction; and atomic weapons, especially hydrogen bombs, to lessen the possibility of war because of fear of mass mutual destruction. As Weinberg conceded,

"Technology will never *replace* social engineering. But technology has provided and will continue to provide to the social engineer broader options, to make intractable social problems less intractable; perhaps, most of all, technology will buy time—that precious commodity that converts violent social revolution into acceptable social evolution."

Weinberg did not define the term "social engineer," but by his use of it he clearly meant the social scientist, who would work hand-in-hand with scientists, engineers, and other technical experts. Far from patronizing social engineers/social scientists, in the manner of Bush, Weinberg praised them: "Social problems are much more complex than are technological problems. It is much harder to identify a social problem than a technological problem." "Quick Technological Fixes"—which thereafter became a popular phrase with both positive and negative connotations—would therefore assist and complement conventional social engineering but would not replace it. Contrary to Weinberg, however, all these "Quick Technological Fixes" were actually not alternatives to, but rather themselves examples of, social engineering.[14]

In their separate ways, Dahl, Lane, and Weinberg, among others, justified the creation of the institutional structures that came to shape federal science and technology policy: the Science Advisory Committee, established by President Harry S. Truman in 1951; its successor, the President's Science Advisory Committee and the special assistant to the president for science and technology, both of which were set up by President Dwight Eisenhower in 1957 following the launching of the first artificial satellite, *Sputnik I*, by the Soviet Union; and the Office of Science and Technology Policy begun by President Gerald Ford in 1976, three years after President Richard Nixon abolished Eisenhower's creations for political reasons.

Eisenhower had already made atomic energy subject to civilian rather than military control. But now politicians, governmental bureaucrats, political scientists, and scientists and engineers could argue that science and technology were no less under the control of the American public than any other realms and that the much-heralded spin-offs to ordinary citizens from nuclear power, spaceflight, and other key areas of federal science and technology funding would indeed benefit everyone in pluralist America.

Equally important, Dahl, Lane, Weinberg, and others tempered the hostility toward politics in technocratic discourse and vision that had previously prevented the Technocrats of the 1930s and like-minded groups from being taken seriously by decision-makers. Where such leading Technocrats as Howard Scott and Harold Loeb had no use for politicians and for politics itself—deeming the whole enterprise messy and morally if not literally corrupt—Dahl, Lane, and Weinberg, along with Galbraith, Bell, and many others, recognized the inevitability of politics, even amid fundamental consensus, and sought to integrate technical and social science expertise with political expertise.[15] They understood that every decision in governance, no matter how seemingly technical, is to some degree a political decision and that the denial of that fact of life by antipolitical Technocrats—and by similarly inclined scientists and engineers—is itself a political stance. As the political theorist Franz Neumann observed, "No society in recorded history has ever been able to dispense with political power. This is as true of liberalism as of absolutism, as true of laissez faire as of an interventionist state."[16] To be sure, the existence of that alleged postwar consensus on national values and vision certainly made the respective positions of Dahl, Lane, Weinberg, Galbraith, and Bell far more palatable than they would have been during the Great Depression, when no such consensus existed, leading to marginal disaffected groups like the Technocrats. And one should not underestimate the skepticism about politics as insufficiently manageable and predictable that is still found among many engineers and scientists.

Still, the fundamental assumption of postwar social scientists like Dahl, Lane, Galbraith, and Bell, and of socially minded engineers like Weinberg, was that public policy both at home and abroad could be managed by the right combination of experts in various fields. This assumption was taken to an extreme by Simon Ramo, among others, who codified it into Systems Analysis or, more accurately, Systems Engineering. Ramo is a cofounder of TRW, a large pioneering high-tech company, and an adviser to several presidents and the Pentagon. Ramo did not alone devise Systems Engineering, but he has advocated, practiced, and written about it more than nearly any other American. Both the modest social engineering at the federal level pioneered by President Hoover and the grander schemes envisioned by Weinberg were dwarfed by the vision of Ramo and his fellow true believers. Systems Engineering was not, however, limited to engineers but instead embraced like-minded social scientists. Moreover, any problems not solvable by the designated original teams of experts would by definition eventually be solved by additions to or replacement of team members, which increasingly meant more social scientists. But any problems not solvable under these conditions were by definition not genuine problems and so could be safely dismissed. Ramo and other proponents of this approach utilized the general systems theory of the 1920s but added a new twist: rather than examine the dynamics of the system or the system itself, these analysts instead focused exclusively on the end product. That focus enabled them to treat any element of any complicated system in a singular fashion: how did it contribute to the output, and how could its contribution to the output be enhanced? Each part of any such system could therefore be looked at as a discrete element rather than, as in earlier systems, a nondetachable part of the whole. Different kinds of experts were thereby needed for each part, and social scientists were thus as invaluable as engineers and scientists. Ramo detailed his scheme in a book entitled *Cure for Chaos: Fresh Solutions to Social Problems through the Systems Approach* (1969).[17]

Ramo's title and subtitle clearly reflected an increasing sense of social, cultural, and political upheaval in the United States and other "advanced" societies in the several years just before the book was published (and, of course, in the first few years immediately after its publication, too). Like the Technocrats of the 1930s, Ramo and his fellow systems engineers had come to see their world as desperately needing the kind of order only they could provide: efficient, honest, nonideological. In the heady atmosphere of the Kennedy and early Johnson administrations there had been little reason to question the logic of Systems Engineering: certainly not until the growing public perception of mismanagement of the Vietnam War by "the best and the brightest" and, no less important, until the campus protests at the University of California at Berkeley and elsewhere that passionately attacked the growing impersonality of education—and, by extension, other spheres of life and work—and the bureaucratic identification of students and workers by numbers and computer punch cards. More precisely, by focusing on output and by assuming that the output is an agreed-upon commodity, systems engineers like Ramo more often distorted than resolved problems. As historian and social critic Theodore Roszak observed, "the good systems team does not include poets, painters, holy men, or social revolutionaries, who, presumably have nothing to contribute to 'real life solutions.'"[18]

Other critics have complained that Systems Engineering not only is elitist, self-justifying, self-perpetuating, and narrowly conceived but also—and most ironically—is inefficient, inadequate, and unscientific.[19] Certainly Systems Engineering hardly enjoys a stellar record in the field in which it was initially embraced: the military. Quite the opposite: a seemingly endless progression of poorly made weapons and vehicles, plus enormous cost overruns, have made many Americans equate military expenditures with waste and greed.

Yet until the mid–1960s the overwhelming public faith in government, on the one hand, and in scientific and technological advance, on the other, had transformed postwar science and technology policy into seemingly progressive social policy about the entire nation, not just the military establishment or the space program or the nuclear power industry. (To this degree, Senator Kilgore might have been pleased.) And with the need to compete with the Soviet Union in science, math, and engineering education following the launching in 1957 of *Sputnik I*, science and technology policy had filtered down to the grass roots more than at any previous time in American history, apart from "hot" wars. What has since been termed scientific and technological literacy became part of daily educational and familial conversations.

Beginning in 1959, with the publication of C. P. Snow's *The Two Cultures*, those conversations were influenced by the notion of a Western academic world bitterly divided into two camps: the sciences and the humanities. (As both a one-time scientist and a highly respected novelist, Snow himself bridged both camps, but he was, by his own account, the exception that proved the rule.) Although Snow blamed each party for ignoring the contributions of the other, he scorned "literary intellectuals" more, for they delighted in their intentional ignorance of science, where scientists more often lamented their limited familiarity with the humanities. Snow also praised scientists who, like himself, had become involved in establishing and administering public policy. Implicit in Snow's plea for greater such involvement by more scientists was the participation of similarly minded social scientists. Like Dahl, Lane, Weinberg, Bell, and other Americans, Snow was quite optimistic about the solution of problems at home and abroad by the application of scientific and, in turn, social scientific methods. He was a fervent, if unofficial, advocate of modernization theory.[20]

But postwar science and technology policy was not just a reaction to the challenge of affluence or to the threat of Communism or to the need for greater scientific and technological literacy. In 1956 William H. Whyte Jr. published *The Organization Man*. Assistant managing editor of *Fortune*, Whyte criticized unthinking, uncritical worship of corporate organizations and pilloried the gospel of scientism, the notion that expert determinations were always nonpartisan and not subject to error or debate. He railed against any organization's use of the concepts of belonging and togetherness to manipulate members and objectives, and he called on his contemporaries to challenge the status quo. Whyte conceded that the organization was here to stay, and he did not advocate nonconformity. Rather, he urged men and women to examine the premises they took for granted, to vent their individual proclivities, and to redirect their various organizations. Individualism tempered by critical thought would awaken America from its blissful organization-inspired ignorance and produce a better tomorrow.

Whyte's book drew plaudits from many corners, became an immediate best-seller, and remained popular for years. Its cry for intraorganizational individualism struck a chord with its readers, as did its attacks on professional, technical expertise as the sole criterion for making decisions. In the 1950s and early 1960s opportunity served as the watchword, as individuals began to assert their interests and proclivities in search of a more fulfilling existence. More than ever before, science and technology became a means to give flight to individual expression, which invariably had material components. Science and technology would give each American the chance to "keep up with the Joneses," the 1950s idealized personification of unfettered opportunity.

In the late 1960s and 1970s, of course, Whyte's anticorporate message would take on new meaning as many young Americans rejected not only the corporate ethos but also the obsession with material goods that went hand-in-hand with it. Yet in some ways, Whyte's greater legacy is the early critique of expertise—and not just in corporations but also throughout American life. Nowhere was expertise taken more for granted than in science and technology and, as already indicated, in the social sciences that emulated them.

Officially, it was President Eisenhower's 1961 farewell address that began the unraveling of the postwar consensus over science and technology policy, a consensus that fell apart over the Vietnam War, the "Star Wars" missile defense system, urban crises, environmental crises, and the higher education–military research nexus—the focus of Eisenhower's concern. By the mid–1960s federal support of science was criticized as inattentive to social issues in a manner reminiscent of Senator Kilgore, but now in a climate of criticism rather than of celebration of American life. In 1968 Congress amended the NSF's chartering legislation to support more formally and more fully applied research and the social sciences, both already funded on a modest basis. The NSF even began an avowedly social scientific program, Research Applied to National Needs (RANN). In the Ronald Reagan, George H. W. Bush, and Bill Clinton administrations, however, socially useful research was redefined as research making high-tech America more competitive vis-à-vis other major powers, Japan above all. Ironically, Senator Kilgore's postwar dream was of a national science and technology policy and structure that would help small businesses, not giant corporations.

Yet federal policy, however supportive of the social sciences, cannot alter the sea change in recent years in public attitudes toward science and technology. The historic bedrock American faith in scientific and technological progress, and in such progress as leading directly to social progress, has clearly diminished. If it is premature to proclaim the end of the Enlightenment Project, much less the "end of science," the evident growth of technological pessimism has made the optimism of the late 1950s and early 1960s seem hopelessly naive and utterly passé.[21] If, for instance, ever fewer citizens care about the space program, so ever fewer are impressed by the spin-offs from it that for so long were used by NASA to justify the program itself. Not surprisingly, the twenty-fifth anniversary of the initial moon landing on July 20, 1969, evoked far less interest than did similar celebrations earlier in American history of significant anniversaries of, say, the opening of the Brooklyn Bridge or the completion of the transcontinental railroad.

Put another way, there has been a steady erosion of faith in recent decades in what psychologist Kenneth Keniston terms the "engineering algorithm"—the core belief that "the relevant world can be defined as a set of problems, each of which can be solved through the application of scientific theorems and mathematical principles."[22] This belief at once underlay and justified postwar science and technology policy and the expanding role of the social sciences in applying scientific and technological advances to social policy at home and abroad. The diminution of that belief has in turn led to ever more radical positions on the nature of science itself. If, as some outside of policy circles now contend, science is socially constructed, any claims of genuine expertise, much less of total objectivity, are logically impossible. And no systems engineering team, no social science models, and no computer programs can change that. This declining national faith in scientific and technological progress has inevitably lessened the complementary faith in the social sciences.

As social critic Kirkpatrick Sale asked in 1980 about the nation's post–World War II track record of "Quick Technological Fixes":

> do we seem to be moving toward real and healthy solutions to our nation's crises, does the populace seem safer and healthier and happier with it all, or do we seem, somehow, to have accumulated problems instead of dispelling them and to have created a world of greater anxiety and risk and chaos than we had before? Solutions, we must remember, are very much like problems: they are rooted in people, not in technology. Schemes that try to devise miracles to bypass people, negate, deny, nullify or minimize people, will not work.[23]

Sale's critique may be excessive, but the repeated failure of the social sciences to provide lasting solutions to so many social problems has certainly contributed to that sea change in public attitudes toward science and technology. As economist Richard Nelson detailed in *The Moon and the Ghetto: An Essay on Public Policy Analysis* (1977), too many social scientists working in science and technology realms have never appreciated the gap between their elegantly rational methodologies and analyses and the messier and more complex conflicts of interest in the "real world" they seek to illuminate and to improve.

The elimination of the Office of Technology Assessment (OTA) in 1995 by the then new Republican congressional majority exemplified that sea change. Not only was OTA officially apolitical, but also its budget was so modest that its shutdown was as much symbolic as substantive. This point was conveniently overlooked by the Republicans who killed OTA in order, they claimed, to demonstrate their willingness to tighten Congress's own belt so as then to justify deeper budget cuts affecting ordinary citizens. Yet the end of OTA also exemplified, contrary to Lane's 1966 analysis, the persistence of ideological politics in the very realm where "universalistic scientific criteria" applied by "professional problem-oriented scientists"—and social scientists—might have been expected to triumph. If anything, science and technology policy are today more political and more ideological in nature than ever before in American history.[24]

Many Republicans perceived OTA as an invention of liberal Democrats. To these Republicans, OTA was anything but nonpartisan in its analyses, contrary to its explicit mission. To them, OTA routinely favored one outcome in its evaluations—that desired by liberal Democrats—rather than simply present all possible options, as OTA was mandated to do. In the process, these Republicans charged, OTA repeatedly misused the social sciences despite a pretense of objectivity and expertise. The new majority's willingness to eliminate OTA was therefore considerably more than a self-righteous budgetary cutback.

Still, it was ironic that the assault on OTA was led by then House Speaker Newt Gingrich. Not only had Gingrich deemed himself a high-tech visionary and admirer of gurus Alvin and Heidi Toffler, but it was Alvin Toffler's best-selling *Future Shock* (1970) that, for all its flaws, made millions of readers aware of the need to anticipate the future more systematically than had ever been attempted. The disappearance of OTA hardly contributed to this effort. That a self-proclaimed leading policy-maker like Gingrich—with a Ph.D. in history to boot—had pushed for its disappearance appeared to make no intellectual sense. And if, in some early cases, OTA was not strictly neutral, in virtually all of its later cases it was avowedly nonpartisan and conscientiously served all its congressional constituents. As political scientist Bruce Bimber puts it, OTA gradually "developed a *strategy of neutrality* . . . not as a professional standard but as a political survival strategy to ward off critics." If anything, OTA kept such a low bureaucratic profile that, according to Bimber, it failed to develop sufficient political support that might have saved it from elimination by persons with political agendas ultimately having little to do with OTA itself, including Gingrich.[25]

Moreover, however partisan OTA might have been perceived as being, it was set up in 1972 not to block scientific and technological developments but instead to assess their potential ramifications, positive and negative, for American society. Its creation reflected the initial wavering American faith in scientific and technological progress, for if there had been no skepticism about progress, there would logically have been no reason to assess those potential ramifications. In the wake of *Future Shock*'s extraordinary popularity, the special concern at the time was preparing for the unexpected—the "effects on all sectors of a society that may occur when a [particular] technology is introduced, extended, or modified, with special emphasis on any impacts that are unintended, indirect, or delayed."[26]

In the early 1970s, though, within both parties there was still sufficient bedrock faith in government and in scientific and social scientific expertise to justify OTA's establishment. If the future was increasingly uncertain, social scientists like Dahl and Lane and socially oriented scientists like Weinberg would nevertheless be able to save the day. Whatever politics lay behind OTA's demise, its elimination did not reflect any lessening of that wavering American faith in scientific and technological progress. Quite the opposite: the subtext of OTA's disappearance was a belated acknowledgment of this sea change.

To be sure, one cannot ignore the resurgence in certain quarters of that faith. David Noble's *Religion of Technology* (1997) illuminates extremes of that faith among those working today in nuclear weapons, spaceflight, artificial intelligence, and genetic engineering. They routinely link their achievements with the recovery of human divinity, lost after Adam's fall. Yet the more telling point is the growing absence of these outdated beliefs among ordinary Americans, where in the past such citizens would happily have gone along with this latest version of the nation's "Manifest Destiny." The foremost example of this same sea change is surely the ambivalence if not outright hostility of most Americans toward ongoing and projected experiments in biotechnology, ranging from the production of "Frankenfoods" to the cloning of plants, animals, and, potentially, human beings.

This brings us back to Vannevar Bush's *Science—The Endless Frontier*. In 1994 the Clinton administration issued a report entitled *Science in the National Interest*. Billed as the first official presidential statement on science policy since 1979 and personally endorsed by both President Bill Clinton and Vice President Al Gore, the report made a passionate case for expanded federal support of science and technology in the post–Cold War era. It was also an avowed sequel to *Science—The Endless Frontier*. The new report's first section was even entitled "Science: The Endless Resource." The only explicit qualification the 1994 report made of its 1945 predecessor was in acknowledging that the "societal benefits" of science and technology derive from an "interdependence" of basic and applied research. "We depart here from the Vannevar Bush canon, which suggested a competition between basic and applied research" for prestige, power, and recognition. Trying to avoid Bush's hierarchy, the 1994 report characterized the contemporary science and technology relationship as "more like an ecosystem than a production line," a production line presumably exemplifying one group (scientists) controlling another (engineers and other technical experts, including social scientists) and demanding conformity by the latter to the former's conception and pace of work—or as some historians have characterized it, the misleading assembly line model of technology in which we put science in and get technology out. That a production line might well be the crowning achievement of many engineers apparently did not occur to the report's writers. Elsewhere in the report, the common conception that Bush perpetuated of engineers merely applying what scientists discover came through, no matter how unintended. The report's very title slighted technology, as did its repeated paeans to the frontiers of scientific discovery but not to invention.

More precisely, one would never have gathered from *Science in the National Interest* that some of the nation's (and the world's) most significant postwar research has been and will continue to be both basic and applied, at once advancing fundamental understandings of science and helping to solve major practical problems. As political scientist Donald Stokes demonstrated in *Pasteur's Quadrant: Basic Science and Technological Innovation* (1997), it is quite possible for first-rate scientific research to be motivated by avowedly practical objectives without sacrificing contributions to scientific knowledge. Contrary to Bush's model, such linkages could enhance, not undermine, basic research, not least for those contemporary policy-makers eager—in a manner akin to Senator Kilgore—to see some fruits of government funding in the form of material improvements.[27]

Stokes rejected Bush's linear model of basic science leading to applied science and in turn to technology and then to commerce and industry in favor of a model using the geometry of "quadrants." According to Stokes, all research falls into one of four quadrants. One quadrant, named for the theoretical physicist Niels Bohr, is for research like Bohr's on the atom that is exclusively basic, never applied. Another quadrant, named for the inventive genius Thomas Edison, is for research that is exclusively applied, never basic. A third quadrant, to which Stokes gives no name, is for research that does not seek to advance either basic or applied science; examples would be nineteenth-century classification projects in natural history. The fourth quadrant, named for the great French chemist and life scientist Louis Pasteur, is for Stokes the most important, for this quadrant encompasses research that, like Pasteur's, is both basic and applied. It advances fundamental understandings while solving significant practical problems. Pasteur's research "was motivated by the very practical objectives of improving industrial processes and public health. It led directly to applications that saved the French silk and wine industries, improved the preservation of wine and beer, and created effective vaccines."[28] But these applications were based on Pasteur's breakthroughs in developing the germ theory of disease and in establishing the field of bacteriology—hence Stokes's endorsement of this model of research for contemporary American scientists, engineers, and, not least, social scientists.

Most important, *Science in the National Interest* ignored the point made by Billington, Vincenti, and other historians of technology that engineering has significant intellectual properties of its own separate from those of science and that the making of things by engineers requires an intellectual discipline as taxing as the making of discoveries by scientists. Significantly, C. P. Snow recognized that engineering is actually a third culture separate from both science and the humanities. In passages of *The Two Cultures* rarely cited, he acknowledged that scientists can be as patronizing toward engineers and other technical experts—including, implicitly, social scientists—as any self-proclaimed humanists. As Snow put it, "Pure scientists have by and large been dim-witted about engineers and applied science. They couldn't get interested. They wouldn't recognise that many of the problems were as intellectually exacting as pure problems, and that many of the solutions were as satisfying and beautiful" as those set forth by scientists. Instead, scientists invariably assumed that "applied science was an occupation for second-rate minds."[29] (Ironically, of course, engineering and technology generally are not, contrary to Snow, simply "applied science.")

Science in the National Interest touted a number of high-tech advances while ignoring any downsides. For example, it cited genetic-engineering experiments without any mention of attendant controversies; the repair of the mirror on the Hubble space telescope that should never have been so flawed in the first place; and the satellite monitoring of the earth's ozone layer that has been damaged by our own handiwork. And the report presumed a continuing—but, in reality, nonexistent—national consensus in favor of increased federal funding for science and technology. "While we cannot foretell the outcome of fundamental research," the report conceded in a revealing comment, "we know from past experience that, in its totality, it consistently leads to dramatically valuable results for humanity." Bush could hardly have disagreed with ahistorical rhetoric like this, rhetoric antedating the wavering faith in progress that led to the creation of OTA in the first place. But the congressional majorities that in 1993 stopped funding the Texas-based Superconducting Super Collider and that in the past few years have nearly stopped funding the space station surely would have disagreed.[30]

Nowhere in *Science in the National Interest*, then, was there any acknowledgment of the legitimation crisis that has begun to color American science and technology policy. Whyte's early critique of scientific and social scientific expertise has by now become commonplace, as has his cry for intraorganizational individualism. In the process, science and technology have steadily been reduced from the nation's principal sources of authority and the objects of unparalleled reverence and deference to those mere means of enhancing individual expression that Whyte identified in their early embodiments. Moreover, contemporary high-tech advances allow a degree of separation and distance from unwanted outside authorities undreamed of when Whyte wrote. Examples abound: from buying and selling goods and securities through the Internet to choosing from among hundreds of television channels through cable or satellite dishes to seeking romance and adventure through personal Web sites. The fact that these and so many more activities can be carried on at home on personal computers makes them all the more appealing to those growing numbers of Americans who expect and demand insulation and protection from the world about them. Not surprisingly, their varied individual political agendas increasingly do not include taxpayer support for megaprojects like super colliders and space stations.

For that matter, whatever appreciation follows for the scientific, technological, and social scientific advances that allow for these new living and working arrangements does not, as in the past, translate into any celebration of science or technology, much less of social science, as wonders in and of themselves. Instead, it is the empowerment of the individual—as seen from each individual's perspective—that is nowadays celebrated. Science, technology, and the social sciences are praised, if at all, for their practical payoffs rather than for their cultural values. This is not exactly the world envisioned by Galbraith, Bell, Dahl, Lane, Weinberg, Snow, or Ramo—or, needless to say, by Vannevar Bush.

Notes

1. Vannevar Bush, *Science—The Endless Frontier: A Report to the President on a Program for Postwar Scientific Research* (1945; reprint, Washington, D.C.: National Science Foundation, 1990), 12.
2. Daniel J. Kevles, "Principles and Politics in Federal R&D Policy, 1945–1990: An Appreciation of the Bush Report," in Bush, *Science*, xi.
3. On Bush's life and work, see G. Pascal Zachary, *Endless Frontier: Vannevar Bush, Engineer of the American Century* (New York: Free Press, 1997).
4. Bush quoted by John Brandl in latter's review of Donald E. Stokes, *Pasteur's Quadrant: Basic Science and Technological Innovation* (Washington, D.C.: Brookings Institution Press, 1997), in *Journal of Policy Analysis and Management* 17 (fall 1998): 734. I refer to the book later in the text.
5. David P. Billington, "In Defense of Engineers," *Wilson Quarterly* 10 (New Year's 1986): 89, 87.
6. See Walter Vincenti, *What Engineers Know and How They Know It: Analytical Studies from Aeronautical History* (Baltimore: Johns Hopkins University Press, 1990).

7. Two relatively recent works recast the origins of the National Science Foundation and the conflict between Bush and Kilgore in more complex and more historical perspective. See Daniel Lee Kleinman, *Politics on the Endless Frontier: Postwar Research Policy in the United States* (Durham, N.C.: Duke University Press, 1995); David M. Hart, *Forged Consensus: Science, Technology, and Economic Policy in the United States, 1921–1953* (Princeton, N.J.: Princeton University Press, 1998). Their scope and analyses are by no means identical, but each is an excellent study.
8. See the many examples of these studies in the self-congratulatory report of Princeton University's Center of International Studies, *A Record of Twenty Years, 1951–1971* (Princeton, N.J.: Woodrow Wilson School of Public and International Affairs, 1971); the report lists no author. See my critique of modernization theory in Howard P. Segal, "Alexis de Tocqueville and the Dilemmas of Modernization," in *Future Imperfect: The Mixed Blessings of Technology in America* (Amherst: University of Massachusetts Press, 1994), 36–48.
9. See Barry D. Karl, "Presidential Planning and Social Science Research: Mr. Hoover's Experts," *Perspectives in American History* 3 (1969): 347–409.
10. See Charles S. Maier, "Between Taylorism and Technocracy: European Ideologies and the Vision of Industrial Productivity in the 1920's," *Journal of Contemporary History* 5 (1970): 27–61.
11. Robert A. Dahl, *A Preface to Democratic Theory* (Chicago: University of Chicago Press, 1956), 137.
12. Robert E. Lane, "The Decline of Politics and Ideology in a Knowledgeable Society," *American Sociological Review* 31 (October 1966): 657–658, 659; emphasis added. See also Robert E. Lane, "The Politics of Consensus in an Age of Affluence," *American Political Science Review* 59 (December 1965): 874–895.
13. Lane, "The Decline of Politics," 661.
14. Alvin M. Weinberg, "Can Technology Replace Social Engineering?" *University of Chicago Magazine* 59 (October 1966): 6–10, reprinted in Albert H. Teich, ed., *Technology and the Future*, 7th ed. (New York: St. Martin's, 1997), 64, 56.
15. See Harold Loeb, *Life in a Technocracy: What It Might Be Like* (1933; reprint, Syracuse, N.Y.: Syracuse University Press, 1996), ch. 4 and new introduction by Howard P. Segal.
16. Franz Neumann, *The Democratic and the Authoritarian State: Essays in Political and Legal Theory*, ed. Herbert Marcuse (New York: Free Press, 1964), 8.
17. See Simon Ramo, *Cure for Chaos: Fresh Solutions to Social Problems through the Systems Approach* (New York: McKay, 1969). See also Simon Ramo, *What's Wrong with Our Technological Society—And How to Fix It* (New York: McGraw-Hill, 1983).
18. Theodore Roszak, *Where the Wasteland Ends: Politics and Transcendence in Postindustrial Society* (Garden City, N.Y.: Doubleday, 1972), 36.
19. See Ida R. Hoos, *Systems Analysis in Public Policy: A Critique*, rev. ed. (Berkeley and Los Angeles: University of California Press, 1983).
20. On the traditional reluctance of engineers to become involved in both politics and public policy, see Howard P. Segal, "The Third Culture: C. P. Snow Revisited," *Technology and Society Magazine* 15 (summer 1996): 29–32. The Technocrats are an obvious exception, but not a complete exception, for two of their principal leaders—Scott and Loeb—were themselves not engineers, as detailed in Loeb, *Life in a Technocracy*, new introduction by Segal.
21. On controversies over science today see, for example, Gerald Holton, *Science and Anti-Science* (Cambridge: Harvard University Press, 1993); Paul R. Gross and Norman Levitt, *Higher Superstition: The Academic Left and Its Quarrels with Science* (Baltimore: Johns Hopkins University Press, 1994); Andrew Ross, ed., *Science Wars* (Durham, N.C.: Duke University Press, 1996); John Horgan, *The End of Science:*

Facing the Limits of Knowledge in the Twilight of the Scientific Age (Reading, Mass.: Addison-Wesley, 1996).

22. Kenneth Keniston, "Trouble in the Temple: The Erosion of the Engineering Algorithm," unpublished paper, 1992.
23. Kirkpatrick Sale, "My Turn: The 'Miracle' of Technofix," *Newsweek*, June 23, 1980, 12.
24. On this development, see Yaron Ezrahi, "Technology and the Illusion of the Escape from Politics," in *Technology, Pessimism, and Postmodernism*, ed. Yaron Ezrahi, Everett Mendelsohn, and Howard P. Segal (1994; reprint, Amherst: University of Massachusetts Press, 1995), 29–37.
25. Bruce Bimber, *The Politics of Expertise in Congress: The Rise and Fall of the Office of Technology Assessment* (Albany: State University of New York Press, 1996), 51. Bimber readily concedes that OTA "was highly politicized in its first half-dozen years of operation. It was widely viewed as dedicated to a narrow set of political interests, and its technical credibility suffered as a result. But OTA evolved over time to be less *politicized*" (20), though some scholars believe it was no less politicized at the end than at the beginning.
26. Edward W. Lawless, *Technology and Social Shock* (New Brunswick, N.J.: Rutgers University Press, 1977), 5. On the attempted application of technology assessment to history, see Howard P. Segal, "Assessing Retrospective Technology Assessment: A Review of the Literature," *Technology in Society* 4 (fall 1982): 231–246.
27. See, for example, U.S. Representative Vernon Ehlers, "Congress, Science, and the Two Cultures," *University of Michigan Research News* 49 (1998): 22–23. Representative Ehlers of Michigan is the first research physicist ever elected to Congress and is a former professor and chair of physics at Calvin College in Grand Rapids, Michigan.
28. M. Granger Morgan, review of *Pasteur's Quadrant: Basic Science and Technological Innovation*, by Donald E. Stokes, *IEEE Spectrum* 36 (January 1999): 12.
29. C. P. Snow, *The Two Cultures* (1959; reprint, Cambridge: Cambridge University Press, 1993), 32.
30. See, for example, Daniel J. Kevles, Preface, 1995, "The Death of the Superconducting Super Collider in the Life of American Physics," in Daniel J. Kevles, *The Physicists: The History of a Scientific Community in Modern America* (1977; reprint, Cambridge: Harvard University Press, 1995).

Part III

Have the Social Sciences Mattered in Washington?

WHETHER THE SOCIAL SCIENCES have mattered once installed in federal politics is in some sense an unanswerable question. Of the future we have no idea whatsoever might happen. And we can hardly know as yet the past; this volume can only make a down payment on a significant contribution, if that. At this point, however, we do have some clues. For the first half of the twentieth century, government institutions employed increasing numbers of social scientists or consulted with them on a variety of problems and issues, domestic and foreign. In the last century's second half, as the federal government expanded to fight World War II and the Cold War, its leaders thought it essential to have policies about science and technology, including the social sciences. Already we have some indications from the essays in Part II that the social sciences were more the creature of politics than the other way around, and that may continue to appear to be the case with these studies. But the entire field cries out for far more investigation than we have had so far.

What is so attractive about Hal Rothman's wide-ranging and well-informed survey of social science and the environment is the common fate of all environmental policies. He makes it clear that the environmental movement (the main subject of his essay) could grow and prosper in the political system for a variety of reasons, not the least of which is that it was doubtless a relief to many establishment politicians of the Left and the Right, a haven from strife over civil rights, affirmative action, juvenile delinquency, drugs in American life, and, not the least, the troubling war in Indochina. For some time the environmental movement seemed almost free, and certainly lacking in a potential for triggering social contention and conflict. Hence the environmental movement was, initially, like a partly inflated balloon, lumpy, without definition or shape, an epidermis that one could literally poke one's finger at without causing distress. But as this "balloon" filled up, as it took more social and political space, and as it acquired dimension, shape, proportion, and, above all, identity, then different interests in the body politic reexamined their positions on its benisons, or lack thereof. And therein lay the difficulty, for as technology and science became, for most Americans, tangible expressions of individual expression, such as songs or pictures made with the new electronic gadgets, too much "interference" with the environment, too much "regulation" became an obstacle to economic development, progress, and even the absolute ownership of property, especially land or industrial factories on certain properties. In other words, individual choice seemed to be the hallmark of the day, what decided what was important, and what mattered the most in such decisions. Powerful private and public interests were involved; these were no mere playthings for those who liked to hug trees, light candles, and wail about the depredations of the environment by soulless corporations over white wine and brie and crackers. Environmentalists, were they to succeed, had to be made of sterner stuff than that.

Child development and early childhood research is definitely an area of social science that in the course of the twentieth century has "gone to" Washington from the child welfare and child study movements of the early twentieth century. These movements' overall legacy was to emphasize early childhood as a key period of the life cycle. In the 1950s and 1960s, as the research and the advocacy for what became known as Head Start got under way, early childhood research and education became a reasonable stable priority for federal programs and funding. As Kristen D. Nawrotzki, Anna Mills Smith, and Maris Vinovskis, point out, all three variants—kindergarten, day care, and Head Start programs—grew in the post–1960s era. What may not be appreciated is why this was so. Early childhood seems like such a formative period of the life cycle, not merely or mainly because the great Sigmund Freud said it but because it fits perfectly into the commonly accepted social philosophy of the formation of character and social competence of most Americans. As the 1940s radio program of personality *Horace Height* repeated with deadly regularity every Sunday, it is always better to build boys than to mend men. In other words, not only is it less expensive and difficult to train youngsters than adults (and there is some truth in that), but proper child nurture enables society to dispense with a large and complex social welfare program for adults, who should be working, paying taxes, and obeying the law. It is probably true, as advocates of Head Start programs insist, that only about 40 percent of those eligible for these programs ever get into one, and then only for a year rather than the three years Head Start's champions claim will make the program work with maximum efficiency. Yet the reason that Head Start is popular with the Left and Right, but never gets full funding, is that its existence permits the

Left to wax poetic that it does exist and should be expanded and the Right to escape the charge that they care not for defenseless little children. Head Start, in particular, and early childhood research and education, more generally, are an almost perfectly balanced political football, useful to the Left and Right to assuage consciences and to reassure skinflints.

And that brings us to two melancholy riffs on the cities and on the poor, which complete the volume. In "The Death of the City: Cultural Individualism, Hyperdiversity, and the Devolution of National Urban Policy," Zane L. Miller paints a startling portrait of half a century and more of the end of national urban policy, a transition from one conception of how society was structured to another. The traditional perspective being replaced was an ideology of group determinism. Groups, not individuals, formed the basic units of American life; individuals acquired their identities and lifestyles from the fact of their membership and participation in group life, just as groups became the groups they were from their history and experience in their social and physical environments. Thus the culture of a group stemmed from social experience in a particular neighborhood, city, region, or nation or from the experience of a group on a social hierarchy, as indicated by race, class, or gender. This deterministic cultural pluralism, as Miller aptly dubs it, held sway until after World War II, when a new perspective slowly came into view, that of cultural individualism. In this new perspective, individual Americans should define how they should conduct, quite literally, all aspects of their lives—which there should be, in Miller's apt phraseology, a "hyperdiversity of choices" in the selection or design of practically all our wants, everything from the cradle to the grave. Hence individual choices, options, and alternatives about everything in life came to shape urban policy.

In the new urban policy, all individuals could choose what they wished, for they were the basic units of society. Such a perspective helped create prejudices against the state, against experts (especially those not deemed worthy), and even against expertise. All elements of the political spectrum came to embrace these bromides and their deeper cultural assumptions. Now individuals, the grass roots, and local government were supposed to be in the driver's seat; no elites need apply. If at first champions of citizen empowerment thought of government as an ally, that was but a temporary romance. Indeed, cultural individualism made impossible any permanent or stable coalitions; constant division and subdivision was the way of all things in policy and politics. Those social scientists who did not adapt to the ensuing chaos, in which some individuals would want to make tons of personal wealth in the city by pushing growth and development and in which others retreated from such selfish materialism and advocated little or no growth to preserve the environment, or in which some other individuals would choose a strongly traditional religious community and family life, as with Protestant fundamentalism or Orthodox Judaism, and others would select to live in communes in which there were copious opportunity for getting high or a roundelay of sexual partners, not to mention free speech and booze, with little or no community responsibility. Everything became, in short, a matter of individual choice, without exception, hesitation, or question. What this led to in urban policy was decentralization in the extreme and no guiding policy save that of cultural individualism. Miller holds out the possibility of an optimistic future. But in any event we notice a general trend parallel to those in other aspects of social science's history as it "went to" Washington: the extreme individuation and fragmentation of American culture and society undercut and reshaped the social sciences. There seemed little threat of self-anointed social scientific elites in the nation's capital taking over the direction of the nation's cities. At best, they adapted to what the grass roots and their representatives, elected or not, wished to have happen.

Individuation and fragmentation seemed also the fate of the social science knowledge and policy that originally defined the national welfare state, as William Graebner argues in the volume's final essay, an essay that concludes in some ways the first essay in the volume, on affirmative action's origins, for affirmative action and welfare came to share the same intellectual roots in the 1950s and 1960s, as both Graebner and Cravens show. Graebner subjects four texts of the times to close analysis—Daniel Patrick Moynihan's *The Negro Family: The Case for National Action* (1965), Charles Murray's *Losing Ground: American Social Policy 1950–1980* (1982), Martin Anderson's *Welfare: The Political Economy of Welfare Reform in the United States* (1978), and David T. Ellwood's *Poor Support: Poverty in the American Family* (1988). As with the cities, and with many of the other events chronicled in this volume's pages, individual choice trumped group analysis; in this particular instance, the therapeutic perspective and the social pathology of the Moynihan argument was replaced by the Anderson-cum-Murray dismissal of the pathological heritage of African Americans and transformation of all blacks, male and female, into mere workers in a marketplace operating without any historical obstacles or handicaps. The end result was what the forty-second president promised, the end of welfare as we know it, or the Personal Responsibility and Work Opportunity Act of 1996. Thus had social science changed from a group determinism to a simple, individualistic, free choice, and rational marketplace in but three or so decades.

More, however, was involved than the attenuation of the social group as a fundamental building block in social science theory and in public policy discourse. A good case could be made that the one extreme—group determinism—was no more helpful than the other—the allegedly rational individual who made calculated and self-interested decisions in a free marketplace of ideas, plans, public policy, and social action. Neither provided much of a contextualized or nuanced portrait, whether that of the grinding social forces of the African American ghetto or the kind of behavior that the game theorists in the social sciences so fatuously proffered. Tools and instruments and methods were one cluster of problems, serious ones, doubtless. There was also the large problem of modernism and postmodernism, of whether the human mind and its social knowledge could be relied on to be the kinds of instruments that would produce truthful accounts of the world. If it were really true, as some postmodernists argued, that there was not one single truth but many different stories, each a political act, then how could there be positive, objective, neutral science? If there were no longer faith in authoritative expertise but faith in all experts, no matter how credulous their claims might be, then of what use was social science in the first instance?

These were certainly difficult questions that led to even more complex and confounding questions and dilemmas about social science and public policy. What to do about the environment, education, the cities, and the safety net of the welfare state, of these and a host of questions and problems throughout the society? Even that sense of doubt, not to say intellectual paralysis, was testimony to the power of the postmodernist mood in contemporary American society and culture.

Chapter 6

Environment, Government and Academe

The Road to NEPA, EPA, and Earth Day

HAL ROTHMAN

PERHAPS NO METEOROLOGICAL EVENT ever had quite the impact on public policy as the large, stagnant mass of noxious air over Birmingham, Alabama, in November 1971. Although there had been numerous similar instances of air pollution in American history, most notably the killer smog in Danora, Pennsylvania, in 1948, no other incident so clearly reflected the rapid change in the values and policies of American society concerning the issue of environmental quality. Much of the federal legislation designed to assure better environmental conditions was already in place, but the response to the events in Birmingham became the litmus test for a new ethic. The public and legal reaction, which included the issuing of a court order that limited production by more than 60 percent at the twenty-three largest industrial companies in Birmingham, ushered in what has been called the "environmental decade."[1]

During the 1970s the combined forces of government and public outcry and insistence created a plethora of environmental legislation that embodied both new and older types of concerns. The Clean Air Act, the establishment of the Environmental Protection Agency, and other similar legislation and policy decisions reflected the renewed interest of an American public concerned with pollution, the quality of life in urban areas, and the long-term health of the nation's physical environment. Legislation such as the Wild Horse and Burro Act of 1971 and the Endangered Species Act of 1973, both of which protected animals from human action and which were controversial almost from the moment of passage, illustrated the optimism of the immediate post–World War II era and the utopianism of the 1960s. These two currents, embodying an increasing unification of pollution issues with historical concerns about wilderness, wild land, and animal issues, were at times contradictory and at others indistinguishable. The consensus they created was underpinned by the work of a number of notable social and natural scientists who played roles in government as well as by public figures who embraced their ideas. This synergy gave people like Rachel Carson, Lynton Caldwell, Garrett Hardin, Paul Ehrlich, and others considerable influence in a time when government policy was fluid, when agency heads and others sought to embrace new ideas that offered solutions to the crises of the age.

As Americans challenged the dominant values of their society in the 1960s, some regarded economic progress as a liability as well as a boon. Concerns for the consequences of human activity were manifest in popular culture and scholarly discourse. These took on many forms, as a number of different currents challenging conventional wisdom took hold in the United States. While the vast majority of these critiques of society were political in nature, a number of very important ones spoke to questions of the status, meaning, and condition of the environment.

Instead of the efficiency that so dominated the scientific conservation of the turn of the twentieth century, Americans developed a new ethic that emphasized the concerns of an affluent, optimistic society that envisioned no limits to its possibilities. Building on the themes of Lyndon B. Johnson's Great Society—such as the eradication of poverty in America for all time envisioned in the War on Poverty—Americans up and down the economic ladder sought to improve their quality of life. Among the ways they defined to accomplish this was to try and make their world safe from chemical hazards and reduce long-term threats to the globe. At the same time, they developed a heightened appreciation for the aesthetic and recreational qualities of the natural world.[2]

Much of this transformative ethos was couched in the language of the time, a form of discourse that openly flouted norms and challenged the philosophies that long dominated American society. Its emphasis on the quality of human experience was new and reflected a degree of self-importance that the self-denying conservationists of earlier in the century shared but would have found surprising. By the 1960s Americans had become obsessed with individualism, individual rights, and personal entitlement instead of the sorts of collective rights and personal obligations the nation's founders envisioned. As environmentalism gathered momentum, it evolved into a form of cultural protest that, alone among such approaches advanced after the beginning of the 1960s, quickly moved to the center of American society. By the end of the decade, addressing the "environmental crisis" had become part of the consensus of mainstream politics.

This transformation began as much social reform in the United States does, with the work of concerned individuals that captured the attention of the public. The 1960s were full of such statements from a range of directions. Four individuals jumped to the front of discourse, while one symbolic image reflected the same fundamental concern: the human race needed to assess its actions more carefully before continuing along the paths it had chosen. Rachel Carson's stunning work of nonfiction, *Silent Spring*; Garrett Hardin's "Tragedy of the Commons," a seminal essay about human behavior in the physical world; Paul Ehrlich's *The Population Bomb*; and Lynton Caldwell, who created the rationale that underpinned the National Environmental Policy Act (NEPA), all influenced the reshaping of national policy. The practices that resulted restructured the relationship between government and individual citizens, making the concept of a regulatory state an intrinsic part of everyday life for nearly everyone who made their living from the land.

Rachel Carson was, in the words of biographer Linda J. Lear, an "improbable revolutionary." Shy and reserved, Carson had grown up in a bedroom community near Pittsburgh in western Pennsylvania during the 1920s. Influenced by her mother, who read widely and loved the outdoors, Carson learned about the natural world from an early age. She and her mother often walked along the Allegheny River near their home, through woods and wetlands, and across river flats. The Pennsylvania College for Women developed the young Carson's precocious appreciation for nature. Carson's mother wanted her daughter to be a writer, and the young woman majored in English for her first two years at school. At the end of her sophomore year, she made a remarkable decision and changed her emphasis to biology; this made her one of only three majoring in the subject at the school. When she graduated with distinction in 1929, Carson received a fellowship to the Woods Hole Marine Biological Laboratory in Massachusetts. She followed it by accepting a small stipend to Johns Hopkins University in Baltimore, where she completed an M.A. in zoology in 1932. Although she planned a doctorate, the combination of the depression of the 1930s and her impecunious family situation prevented her from achieving this goal.[3]

In 1935 Carson entered the federal service after her father's sudden death made her the family's primary breadwinner. She found a part-time job in the Bureau of Fisheries in the Department of Commerce, where she wrote the scripts for a radio series entitled *Romance Under the Water*. After recording the highest score in the nation on the civil service examination, she received a permanent federal position as a junior aquatic biologist in the Fish and Wildlife Service of the Department of the Interior. Thirteen years later, after having proved that she could explain science in graceful and clear writing the public could easily understand, Carson was appointed editor in chief of all Department of the Interior publications.

In her free time, Carson pursued a private writing career. She first published in the *Atlantic Monthly* magazine in 1937, and her first book, *Under the Sea-Wind*, debuted to public acclaim in 1941. Her second book, *The Sea Around Us*, was serialized in the *New Yorker*, became available in 1951, and was an instant success. It remained on the *New York Times* best-seller list for a number of months and was a primary selection of the Book-of-the-Month Club. The success of the book allowed her to retire from government service and pursue full-time writing.

Since arriving at the Department of the Interior almost twenty years earlier, Carson had been concerned with the application of synthetic pesticides to crops, rangelands, and other habitat. World War II was a watershed in the development of synthetic chemicals. Organic materials commonly used for control of noxious plants were replaced with inorganic compounds such as DDT (dichloro-diphenyl-trichloroethane), a potent chlorinated-hydrocarbon insecticide first synthesized in 1943. Another major type of new pesticide, phosphorous insecticides such as malathion and parathion, also became popular. By 1947 the United States produced more than 125 million pounds of such widely used compounds each year, and the rates of production and use continued to skyrocket. Traces of these new chemicals became pervasive in almost every analysis of the condition of the environment. By the mid–1950s, when Carson received a letter from her friend Olga Owens Huckins that crystallized her nagging concerns, the consequences of the widespread use of synthetics were quite real. One obvious result, the disappearance of songbirds of which Huckins complained, captured Carson's attention and then that of the nation. The title of her book, *Silent Spring*, succinctly captured that fear.[4]

Carson was reclusive, but the serialization of *Silent Spring* in the *New Yorker* in 1962 made her a celebrity and created a national sensation. Indicting the chemical industry, agribusiness, and the federal government for the indiscriminate use of chemicals, the published excerpts from the book inspired tumultuous response from every corner of the "respectable" world of industry science. *Time* magazine accused Carson of an "emotional and inaccurate outburst" in an article about the book, and she was criticized by Ezra Taft Benson, formerly secretary of agriculture and later president of the Church of Jesus Christ of Latter Day Saints. Carson's status as an unmarried woman provoked some fire, while others such as Benson dismissed her in the categorical terms of the time as a "hysterical female."[5]

By the time the book debuted in September 1962, two camps that opposed her work had solidified. One, comprised of Department of Agriculture officials, agribusiness, and the chemical industry, saw the book as a public-relations problem, albeit a nightmarish and expensive one. The other, the members of the professional science community, regarded Carson's work as a challenge to their integrity. What Carson condemned as a "Neanderthal age of biology" was the dominant paradigm in insect control, the source of the income of many industry scientists, the funding of their research, the significance of their profession, and, for many, their self-esteem.

Carson touched a raw nerve in American society. She presented only the case against synthetic pesticides instead of weighing the advantages as well, but that formulation spoke to concerned citizens across the nation. President John F. Kennedy pledged an investigation of the charges Carson's work contained even before the book was available in stores and directed his science adviser to set up a committee to look into the questions she raised. When *Silent Spring* debuted in September 1962, the book was an instant best-seller, with more than 600,000 copies in circulation the first year. Despite an industry-sponsored campaign to discourage the project, the following spring CBS aired a prime-time special entitled "The Silent Spring of Rachel Carson." The toxicity of progress had become a national issue.

Carson's axiom that "we know not what harm we face" spoke to an important legacy of technological innovation in American society. When the United States used the atomic bomb against Japan at the end of World War II, the American public and even the scientific community responsible for it reacted with a combination of awe, fear, and horror. The atomic bomb was an exceptional device, but the more people thought about its implications, the less comfortable with it they became. Individuals as diverse as J. Robert Oppenheimer, who directed the Los Alamos project that developed the bomb, statesman John Foster Dulles, and imprisoned Nazi war criminal Hermann Göring all expressed reservations about its use.[6]

The atomic bomb shocked Americans and the world, ushering in a new age and requiring a kind of reflection about the nature of warfare that previously was unnecessary. Progress acquired a tarnish that it never entirely shed, but that was buried during the dark days of the Cold War during the 1950s, when national security became paramount. Yet that current remained in American society, hidden behind the conformity that marked the 1950s. Rachel Carson's presentation of the invisible consequences of pesticide use spoke strongly to suppressed fears in American society.

In 1968 biologist Paul Ehrlich added another perspective with a book entitled *The Population Bomb*. Building on an argument about the rate of population growth compared to that of the expansion of food production first offered in 1798 by English philosopher and cleric Thomas Malthus, Ehrlich pointed out that humanity was on a course to self-destruction as a result of the sheer weight of its natural increase. In the most famous section of the book, he described how he came to understand this point "emotionally one stinking hot night in Delhi," India. In the greater than 100-degree heat, amid the smoke and the dust, Ehrlich wrote, "the streets seemed alive with people. People eating, people washing, people sleeping. People visiting, arguing, and screaming. People thrusting their hands through [our] taxi window, begging. People defecating and urinating. People clinging to buses. People herding animals. People, people, people, people." Believing that he knew the feel of overpopulation from that experience, Ehrlich sought to alert Americans to the inherent problems of the exponential population growth of the modern world.[7]

Another sensation, *The Population Bomb* sold more than three million copies and kicked off an immense debate about the virtues of having more people on the planet. Criticized as a prophet of gloom and doom, Ehrlich initiated a debate that crossed cultural and religious boundaries, affected social policy, and pointed out the inherent problems associated with unbridled technological advancement. He showed that sixty Indian babies born the night of his visit to Delhi would consume about as much as one American baby who entered the world at the same time. This raised the question of the equity of resource distribution, as well as the problems inherent as the human race engineered solutions to problems of infant mortality, disease, crop production, and other maladies that had long affected the species.[8]

Ehrlich's position challenged the reigning views of the Roman Catholic Church as well as everyone who believed in programs such as the green revolution, which transferred American seeds, technology, and pesticides to the people of the Third World in an effort to help them produce crops for market. Church doctrine in the 1960s opposed any form of artificial contraception, but methods of birth control—the most revolutionary of which was the birth control pill, first made available in 1960—were already common in the Western world. Americans remained notoriously laggard in considering the implications of their own population growth, preferring to regard overpopulation as a function of poverty rather than of resource consumption. Americans could afford babies, many seemed to be saying; it was the rest of the world that could not.

The green revolution—the package of innovations in practice and technologies to implement them designed to increase commercial agriculture production in the underdeveloped portions of the globe—seemed to offer a way to give the rest of the world that capability. Initiated in the aftermath of World War II, the green revolution was designed to bring Third World countries into the global economy by providing them with marketable commodities. Typically the implementation of green revolution programs promulgated hybridized seeds that responded to the new classes of synthetic fertilizers; farm management practices based on use of biocides; and agricultural equipment and machinery to support both extensive and intensive expansion of farming. Often, these packages involved both new irrigation developments and expansion of rain-fed agriculture. In some ways, the process worked. By the late 1960s agriculture had begun to take on market characteristics in many places around the globe.[9]

Mexico offered a prime example. Between 1940 and 1965 agricultural output in Mexico increased by a multiple of four. As a result, that country grew enough to feed its burgeoning population and became a net exporter of food for the first time in its history. Mexican agriculture became a model for the Rockefeller Foundation, the U.S. Agency for International Development, the World Bank, and the other institutions that funded technological advancement and economic development in the Third World. Agricultural technology seemed to offer the solution to the predicament of the human race.

But the triumph of the green revolution was short-lived. By the 1970s most advances reversed. Mexico became a net importer of food, and its agriculture was widely perceived as being in a state of crisis. Crops with great market value, particularly specialty fruits and vegetables such as strawberries, asparagus, and broccoli, replaced staples such as grains and beans in the fields of Mexico. The result was agriculture directed at the international market instead of food production for home, with disastrous consequences for the multitude of Mexico's poor. The imported staples necessary to feed the population were more expensive than homegrown crops, taking a terrible toll on the impoverished.

The green revolution focused on market agriculture, and subsistence production did not undergo a similar revolution. Agricultural land belonged to the well-off in Mexico, and subsistence production was not sufficient to make up the gap. The poor became poorer and migrated in large numbers, first to Mexican cities and later across the border, often as illegal immigrants, into the United States. The rich in Mexico maintained control of the land and its wealth, benefiting economically from the burgeoning market in specialty crops.

The results of the green revolution confirmed the axioms of another of the major environmental thinkers of the 1960s, biologist Garrett Hardin. In 1968, in a presidential address to the Pacific Division of the American Association for the Advancement of Science entitled "The Tragedy of the Commons," Hardin articulated an idea that has remained revolutionary in U.S. society: that there were classes of problems to which no technical solution existed. As a result, he suggested, Americans needed to reexamine their individual freedoms to see which ones were defensible in light of burgeoning social needs.[10]

To illustrate this point, Hardin demonstrated the problems of societies with shared common resources. Individuals maximize their use of common resources, he argued, saving their own until the common resources are depleted. There was only one Yosemite Valley, to which at the time everyone had access at any time; this openness led to an erosion of the values that most sought within the park, a feeling familiar to almost anyone who has thought they have a beautiful place to themselves only to hear the voices of approaching strangers. Common areas, he argued, were justified only where populations were not dense. They functioned well as long as population remained in a stasis, but as soon as substantial growth occurred, for whatever reason, the protection of the commons dissipated. In the thickly populated modern world, Hardin insisted, people needed to make more rules to govern their activities. Greater freedom for all would result from this sacrifice of individual liberty.[11]

In many ways, this argument ran counter not only to the trends of the 1960s but also to the idea of freedom as Americans understood it. Hardin had been concerned with the right to reproduce, arguing that "mutual coercion, mutually agreed upon," was the best way to solve the problems of population and, by extension, those of the physical environment at large. In effect, he argued for an earlier, somewhat archaic definition of freedom as collective rights with personal obligations. The cultural currents of the 1960s favored individual freedoms; it was in the second half of the twentieth century that the right to do what one pleased—wherever, whenever, however, and with whomever—came to be considered a basic American and even human right. Student protests and alternative cultural expressions had much to do with individualism. The famed expression of 1960s angst, "do your own thing," clearly reflected the primacy of the individual. At the time when large numbers of Americans expressed genuine distrust of their government, Hardin advocated commonly agreed upon solutions in the form of rules and laws—in other words, government—to solve the social problems stemming from overpopulation. As did much else during the 1960s, the problems and the best solutions to them seemed paradoxically and diametrically opposed.

Observing the 1960s in the United States, it would have been hard to predict success for Hardin's strategy. He was a biologist, thinking in the empirical terms of science and arguing for common solutions implemented by government at a time when such a remedy to social ills seemed anachronistic at best and more likely misguided. The great outpouring of "antiestablishment" thinking and the destruction of old social, racial, and cultural barriers foretold a more open rather than a more restrictive world. Figuratively freed from their psychic chains, Americans sought more for themselves. Defined in different ways, this thinking permeated the nation and made pleas such as Hardin's audible only in selective quarters.

It fell to Lynton K. Caldwell, a political scientist at the University of Indiana, to pull together the loose ideas floating about and to organize them into a strategy that led to a comprehensive policy. In *Environment: A Challenge to Modern Society*, published in 1970, Caldwell linked the social thoughts of natural scientists with the policy experience of a political scientist. Advocating "intelligent and moral action"—that is, policy designed by experts in social affairs—Caldwell predicted that the survival of the human species depended on an ethos not dissimilar to Hardin's, a form of mutually agreed upon mutual coercion. Caldwell regarded "environmental administration," a phrase that surely was an oxymoron less than a decade before, as a significant step. He placed far greater emphasis on "the control of human action in relation to the environment" than on any other aspect of the subject. "The actions of people," Caldwell wrote, "as they impinge upon the environment become the direct focus of attention."[12]

Caldwell's work was rare in its ability to synthesize a range of ideas into a readable volume and even more rare in both its impact and its policy implications. Caldwell's proscription, a managed society, was not new. It echoed the Progressive Era conservationists as well as people such as Garrett Hardin and had been widely applied in post–World War II America. What gave Caldwell's message powerful cachet was its definition of a way to manage the environment through the creation of a new government agency, a Department of the Environment. While in some ways this ran counter to the antigovernment trends of the time, Caldwell softened the blow by arguing for environmental management as an applied science, a public function, and an ethical system. In this he embraced the regulatory role of government but tempered it with the primary influences of the time. Caldwell translated needs and desires into a public policy prospectus, laying the onus on government to administer and to be fair-minded at the same time.

Caldwell's proscription built on almost a generation of individual pieces of legislation as well as on the intellectual basis of the cultural revolution of the 1960s. Nothing prepared Americans for the expansion of the federal role after World War II. By the 1960s environmental regulations were near the top of the list of changes that had occurred in everyday American life. New categories of rules and laws, aimed at making the shared areas of American life—the roads, parks, factories, and neighborhoods of the nation—more habitable were added to the lists of regulations for federal land. Much of the impetus for this transformation followed 1960, in many ways part and parcel of the larger cultural revolutions that took place in American society.

Quality-of-life issues especially reflected the changing nature of postwar American life. During the 1960s the United States was a particularly optimistic society, willing to undertake challenges such as disease and poverty in its midst. Young Americans especially developed strong utopian tendencies, seeing the promise of America as unfulfilled and the world as perfectible by their efforts. A cleaner environment—a response to the gross fouling of the continent that was linked to the great burst of economic prosperity that followed World War II—seemed a reasonable goal for a powerful, affluent society. Americans had the wealth and the desire to think as much about the future shape of their nation as of its present condition, a situation that spawned much of the optimism that was evident at every level of U.S. society. In an age when science retained a reputation for fair-minded objectivity, social and natural scientists became the linchpin of that sentiment.

It was easy to contrast the social scientists of the 1950s with their counterparts who followed. In the 1950s scholars and thinkers such as William Whyte, Betty Friedan, Daniel Bell, and David Riesman enjoyed major influence on public thinking and ultimately government policy. But their concerns did not include the environment; instead, they focused on the human condition in an industrial society, stipulating to the state of the environment as a baseline from which to begin. Their successors applied their intellect and skills to the impact of human beings on the environment and the consequences for life on the planet. The result was a revolution in American attitudes, one that in the end questioned the ideal of progress and, in some instances, the prerogative of humanity on the planet. That changed emphasis played out all over the American landscape but especially in the enactment of laws that regulated human ability to transform the physical world.

A federal presence in the new environmentalism emerged almost overnight. In 1968 the Brookings Institution did not rate the environment among the important issues facing the nation; the following year, Senators Gaylord Nelson of Wisconsin and Edmund Muskie of Maine took the lead in Congress, and a reluctant President Richard M. Nixon signed the landmark National Environmental Policy Act of 1969 (NEPA) after it sailed through Congress. The massive support for the measure indicated both its great importance and the consensus concerning its necessity; only the most significant and the most widely popular legislative proposals generated such widespread support.

NEPA reflected the environment's importance to the voters in American society. The environmental crisis as the issue was called became a buzzword with nearly universal recognition. Americans believed that the protection of their environment was a "good," a socially advantageous and highly desirable goal. The passage of NEPA reflected the public's feelings. It committed the government to "create and maintain conditions under which man and nature can exist in productive harmony, and fulfill the social, economic, and other requirements of present and future generations of Americans." The set of obligations included in this language reflected the dramatic shift in the meaning of the environment to the American public. Front and center, environmental quality had become a standard of measurement in American life.

The Founding of the EPA

In December 1970 the signature event of the environmental revolution—the establishment of the Environmental Protection Agency (EPA)—took place. Established under the auspices of NEPA, the EPA was the centerpiece of the emerging federal environmental regulatory system. The new agency was supposed to centralize the diverse parts of the federal bureaucracy that had responsibility for the administration of environmental affairs. The EPA supplanted agencies such as the Federal Water Quality Administration in the Department of the Interior, which previously handled water pollution enforcement, and the National Air Pollution Control Administration, located in the Department of Health, Education, and Welfare. It also administered the many solid-waste management programs scattered throughout the government; set standards and guidelines for radiation control, a task formerly the province of the Federal Radiation Council; and handled pesticide and toxic substance registration and administration. Created as a "line agency," one with a budget line all its own, a $2.5 billion budget by 1972, and more than 7,000 employees, the EPA was designed to wield a powerful club on behalf of the bipartisan coalition that drove American environmental politics.

With the establishment of the EPA, the federal government codified the era's new environmental ethic into both policy and the legal code. Prior to the early 1970s, economic justification was sufficient for any private or public undertaking. Urban renewal, which demolished much of the historic fabric of American cities, proceeded without statutory regulation until 1966, when the Historic Preservation Act was passed. Despite the Clean Air and Clean Water Acts of the 1960s, pollution continued nearly unabated, and air quality continued to worsen. Dams and other public works projects were built with the flimsiest of rationale and little concern for either their economic or environmental consequences. The creation of the EPA sent a different message to the public, one suggesting that new methods of operation would redefine the patterns of administration that governed the use of the American environment. This powerful new agency, with its many ways to compel compliance, signaled the advent of a world in which the federal government and private industry would assess the consequences of their actions in advance of initiating them and would consider alternatives to any damage that might be caused. It was a new era for Americans. The federal government assumed a greater level of responsibility for environmental conditions than previously had been expected or required by law.

Earth Day

The first Earth Day, which occurred on April 22, 1970, exemplified public response to the idea of a quality-of-life-based environmentalism. Reflecting both the independent spirit of 1960s politics and the social healing essential to the process of co-opting ideas from the American cultural revolution, Earth Day started inauspiciously as a series of environmental teach-ins, formally informal sessions in which people expressed their views, at colleges, high schools, and community centers across the country. From these meager beginnings it metamorphosed into a major American cultural event.

Gaylord Nelson, the U.S. senator from Wisconsin and an outspoken advocate of environmental quality, first developed the idea of a "national teach-in on the crisis of the environment" at a September 1969 symposium in Seattle. Nelson's objectives were twofold: to help crystallize the environmental constituency and to limit its links with the New Left so prominent on American college campuses at the time. Nelson believed that the environmental crisis, as the subject of environmental quality was then called, was the greatest dilemma facing the human race, one that transcended social, cultural, political, economic, and geographic boundaries. The teach-in he envisioned would begin the process of alerting everyone to the subject's significance.

By the end of 1969 the idea of the teach-in had generated a great deal of attention. Nelson's office remained at the center of the issue, and a new organization, initially named Environmental Teach-in Inc., was founded to handle queries about the idea. Environmental activists from across the full range of the political spectrum became involved as the idea of the teach-in gained currency. "Once we announced the teach-in, it began to be carried by its own momentum," Nelson remembered. "If we had actually been responsible for making the event happen, it might have taken several years and millions of dollars to pull it off. In the end, Earth Day became its own event."

In the hands of its chief organizer, Harvard University law student Denis Hayes, Earth Day became a centrist event. Hayes eschewed the confrontational politics of the New Left, seeking instead to unite the supporters of environmental issues rather than polarize them. "We didn't want to alienate the middle class," Hayes remarked later. "We didn't want to lose the 'Silent Majority' just because of style issues." Hayes and the other organizers hoped that the decentralized nature of the event would allow it to avoid the confrontational stance of so much of American public life during the late 1960s. The organizers fashioned a celebration as much as a critique of American society, a search for consensus as well as alternatives. Although this gave Earth Day the widest possible reach, it cut into support among some of the most activist constituencies. These groups equated militancy and activism and found the tempered rhetoric of Earth Day a little tame.

Earth Day was a rousing success. As many as twenty million people around the nation participated. The advocacy group that grew out of the teach-in, Environmental Action, described Earth Day as the largest, cleanest, most peaceful demonstration in American history. Hayes reflected the sentiments of many of the people involved: "We will not appeal any more to the conscience of institutions because institutions have no conscience. If we want them to do what is right, we must make them do what is right. We will use proxy fights, lawsuits, demonstrations, research, boycotts, ballots—whatever it takes. This may be our last chance." Mixing 1960s ideology and the rhetoric of moral suasion with the tactics of the civil rights movement, Hayes fashioned a strategy that simultaneously included and excluded radicalism. In the environmental movement, he seemed to say, everyone had a stake, no matter what their politics.

The main focus of Earth Day became a consensus-oriented form of education. The typical program included a convocation, singing, dancing, food, and rhetoric from environmental advocates. There was also a political dimension; petitions advocating the cessation of local industrial activities that polluted were sometimes circulated, and political candidates who espoused pro-environment values used the event as a stage. The educational effort was also directed at a younger audience in the elementary school grades. In one instance repeated across the country, a sixth-grade class marched outside on a raw April Illinois day to plant a tree on its playground as a way of participating in the celebration.

This inclusiveness gave the movement wide currency in the nation as the public awoke to the idea of the environment as a social issue. Environmentalism began to compete for the mantle of secular religion in modern American life. Its symbols caught on widely, as its message spread throughout the counterculture and the establishment. Americans from all realms embraced the values of environmentalism; politicians and "longhairs" together talked of the same goals. Legislatures followed with supportive action, including the passage of bottle bills that mandated the return of glass for reuse and other similar laws. For a brief moment, it seemed as if the broad spectrum of support for environmental quality would heal the immense political rift in American society. Here was an issue that a wide range of Americans, from the leaders of industry to the political left, could embrace as a social objective.

The Limits of 1970s Environmentalism

But this positive sentiment proved illusory. Although environmentalism generated a great amount of public support and managed to find a home in bipartisan politics in the 1970s, it failed to reach into every corner of American society. Centrism meant many potential adherents, but it also watered down the intensity of environmental support. The large numbers of people interested in the topic suggested environmentalism attained a prominent position in American society, but their interest was not always very deep. Nor did mainstream environmentalism reflect the breadth of the American spectrum. Conspicuous by their absence were minorities—African Americans, Hispanics, and Native Americans, in particular—and people from rural areas. Conservationists and environmentalists historically had come from the classes of people who were economically secure, the ones who went to college and were exposed to the written culture of the United States. Environmentalism's message about deferring material gain in order to preserve the future held little appeal for the poor or others previously excluded from economic prosperity. Farmers, increasingly dependent on chemical technologies for greater crop yields in the nearly century-old effort to combat the problems of surplus with more surplus, and ranchers, embroiled in a constant struggle to maximize the efficiency of marginal public and private grazing land, were left out of what was essentially an urban-based vision.

Environmentalism in the 1970s dealt with issues such as pollution that affected everyone, but often its language and form of presentation spoke to the feelings of city people about distance from the natural world. Despite the emphasis on regulation embodied in the thinkers who underpinned environmentalism and legislation such as NEPA, environmentalism retained an ideology focused on the outdoor issues that had been the province of the Sierra Club and the Wilderness Society, even as pollution and urban issues became the focus of public interest and the legislation that inevitably followed.

One result of this influence was the Endangered Species Act of 1973. Species extinction had been one of the consequences of the attitudes of Americans toward wildlife. A recurring theme in American literature since James

Fenimore Cooper wrote during the 1820s and 1830s, the wanton destruction of wildlife accompanied the Euro-American march across the continent. A range of species, including the most famous, the passenger pigeon that once crowded American skies, were eliminated, and numerous other species teetered on the brink of oblivion. Even the ubiquitous prairie dog, once seemingly everywhere, was reduced to a relict population.

Legislation to protect the most threatened species was first passed during the middle of the 1960s. The Endangered Species Preservation Act of 1966 was another of the prototypes for environmental legislation. It provided for the protection and propagation of native species of fish and wildlife, from the largest vertebrates to even the minuscule Devil's Hole pupfish, a fish less than two inches long that lived in one sinkhole in Ash Meadows, Nevada. A subsequent bill, the Endangered Species Conservation Act of 1969, expanded protection from vertebrates to mollusks and crustaceans as it extended the range of species covered from only those endangered to those threatened by humanity. The Endangered Species Act of 1973 became the final stage in the process of developing protective legislation. This bill reaffirmed the importance of habitat protection, although not without controversy. The 1973 act focused on the protection of "critical habitat" but did not define "critical" in any sort of legal or administrative context. A battle over semantics resulted, in which economic and environmental needs were uncomfortably juxtaposed.

The Endangered Species Act made national headlines during construction of the Tellico Dam on the Little Tennessee River in the mid–1970s. When University of Tennessee zoology professor David Etnier discovered a small fish called the snail darter in the Little Tennessee River in the area that would be flooded behind the nearly completed $116 million dam project, a test of the ethic of bipartisan environmentalism began. A two-year legal battle over continuation of dam construction ensued. Although the Endangered Species Act was widely blamed for the cessation of work on the project, an interagency committee found that the dam was also a classic piece of pork-barreling, a project with no economic feasibility other than to provide jobs in the district of a powerful congressional representative.

An Energized Public

In the early 1970s issues such as wilderness continued to inspire the greatest response from the American public. Wilderness had an iconographic meaning that gave it millions of armchair supporters, and changing technology gave many more people the opportunity to seek wilderness experiences. The combination of easier mobility along interstate highways, better equipment and technology for outdoor activities, and a heightened sense that quality of life included issues such as pitting oneself against nature and learning to appreciate the natural world created a revolution in American recreation. With lightweight gear, more comfortable clothing and footwear, and better-tasting freeze-dried food, wilderness in particular and the outdoors in general became fashionable.

Americans translated their growing interest in the outdoors into involvement in environmental organizations. Many environmental groups doubled or tripled in size in a brief period. A vociferous segment of the public—energized by the enactment of legislation, confident in a value system that challenged authority, and suspicious of the pronouncements of government in the aftermath of the Johnson administration's loss of credibility over Vietnam and the fall of Nixon in the Watergate scandal—began to play a much more significant role in questions concerning wilderness as well as in environmental affairs in general.

The range of legislation passed in the first years of the environmental decade facilitated public involvement in environmental issues. The spate of new bills covered nearly every possible aspect of environmental management, from the Wild Horse and Burro Act of 1971 to the Fishery Conservation and Management Act of 1976. The environmental impact statement process required publication of draft documents and public comment on the proposals. Development soon faced challenges not only from local people directly affected by such planning but also from a national constituency that might complain about federal decisions. Despite the fact that it was sometimes flawed, the comment process allowed and even encouraged a level of public participation that had not previously existed. It energized the public, made it feel a part of decision-making, and gave environmental issues a public constituency and a widened audience.

This new situation created severe management problems for federal agencies. Weakened by their bouts with an energized public and seeking to please the loudest voices around them, federal agencies ran up against the management problems inherent in the issue of wilderness. In case after case, federal administrative decisions that evaluated areas covered by the description of potential wilderness in the Wilderness Act were challenged by a range of interest groups. In New Mexico, a National Park Service recommendation of "no wilderness" at Bandelier National Monument was defeated by a vociferous public that included people from across the country. Ironically, a significant number of supporters of the wilderness revealed that they did not understand the parameters of legally designated wilderness when they supported it; they advocated the inclusion of developed areas within the boundaries of the designated wilderness, a clear contravention of the Wilderness Act. On the Navajo reservation, challenges to the Peabody Coal Company's environmental impact statement for a slurry pipeline to carry coal eventually led to an increase from the pre–1970 price of $5 per acre-foot for water to a more realistic assessment of the impact of the one-way pipeline off the reservation, a cost of $600 per acre-foot implemented in 1987. The Forest Service faced repeated attacks on its wilderness policy, as agency officials tried to maximize the timber cut from national forests at the same time that segments of the public sought to preserve as much roadless land in designated wilderness areas as possible. In the early 1970s frustrated foresters developed a comprehensive review process, the Roadless Area Review and Evaluation (RARE), in response to the situation.

After 1972, RARE proceeded rapidly. Areas of roadless forested land throughout the country were evaluated for their appropriateness as wilderness areas. The Forest Service adopted a three-alternative system: an inventory produced areas that the agency determined were suitable for wilderness and others that its reviewers thought inappropriate. The places that remained were slated for further study. The agency was caught between powerful special interests. The timber industry became incensed when what its members regarded as valuable timber was locked up in Forest Service proposals to designate wilderness, while wilderness proponents believed too many roads were planned and too much timber was to be cut. The tension became particularly fierce over lands for which further study was planned. Wilderness proponents especially saw this intermediate category as a prelude to cutting, fearing the political power of the timber industry and its strong ties to Congress. Timber companies did not understand the sentiments that opposed their position and regarded their opponents as merely and unjustly antibusiness. Weakened by the ramifications of the political and cultural climate of the 1960s, which increased the vulnerability of federal agencies, the Forest Service vacillated. RARE became unpopular across the board. No one was happy with the results of the process.

Forest Service officials determined to try again. Assistant Secretary of Agriculture Rupert M. Cutler and U.S. Forester John R. McGuire initiated a second roadless area review program, called RARE II, which sought to include the widest possible public participation. Again good intentions blew up in the agency's face; the public meetings foresters so wanted became tumultuous and contentious as the changing economic climate, burdened by post–Vietnam War inflation and the increase in fuel prices caused by the OPEC oil embargo, cut the margin that separated success and failure for timber companies and ranchers to a minimum.

The cries of the new environmental coalition a few years before had not seemed threatening. But by the mid–1970s everyone in a range of industries felt that entire ways of living and earning a livelihood were on the line. Too much resentment and not enough constructive input about the selection process followed. The range of special interests that demanded their piece of the Forest Service pie seemed even larger and more unmanageable than ever before. Finding a middle ground among the various constituencies became even more difficult, as the Forest Service, weakened by the changing political climate and new rules governing the behavior of federal agencies, sought to serve disparate and usually mutually exclusive interests.

Part of the problem the Forest Service faced was its long-standing emphasis on commercial economic uses of national forest land at a time when a large segment of the American public espoused a more holistic approach to the use of natural resources. Tied to the ethos of Progressivism, the Forest Service, at its core, still believed in regulated management of natural resources, the principle articulated by Gifford Pinchot more than seventy years earlier. Pinchot had brilliantly shaped public opinion; his successors reacted to the demands of the environmental public and the timber industry, both of which sought to mold agency policy to their objectives. Meeting the demands of this new political and cultural setting, the heightened environmental awareness and the more strident demands of industry, meant throwing over the basic tenets of the agency. It was a step that foresters were not prepared to take.

The result squeezed the Forest Service between the powerful and vocal environmental public on one side and timber companies on the other. After 1960, when it enacted the doctrine of multiple use, the Forest Service had straddled the line between the increasingly active public and the industries that had been instrumental parts of its base of support. The increase in recreational use of national forests and the growing primacy of outdoor experience in American culture forced the agency to make choices it had always hoped to avoid.

When the agency selected only fifteen million acres of the recommended thirty-six million acres to be preserved as wilderness under RARE II, all sides exploded in fury. Traditional wilderness and timber industry adversaries protested at what they described as the destruction of their interests. The recreation industry split; some factions advocated wilderness as the most efficacious and profitable approach, while others saw development as the basis for growth of the industry. Hunters and off-road vehicle enthusiasts attacked designated wilderness as a threat to their interests. In the gale-force winds that resulted, the Forest Service tried to hold on to its position.

Environmentalism and New Economic Realities

The multifaceted debate over the RARE II program, the reauthorization of the Endangered Species Act, and a host of other environmental situations reflected the changing tone of U.S. life during the late 1970s. As long as the American economy continued to expand, or at least the perception that opportunity was growing was commonly accepted, the removal of resources from potential production was a possibility. Conditioned by the economic successes of the 1950s and 1960s, Congress became accustomed to, in effect, apportioning the pie that was the American economy. Most resources were directed toward production, but a few, particularly spectacular scenery and remote places, could be designated for spiritual, recreational, and other not inherently or evidently profitable uses. It was the mark of a "civilized" and mature society that understood its technological limits, many thought, evidence of a compassionate and rational culture that could also advocate ideas such as the end of poverty in America for all time.

But after the OPEC oil embargo and in the middle of Vietnam era–inspired inflation, the nation's belief in ever-expanding prosperity began to wane. Coupled with the increase in expectations across the nation that accompanied the great economic aberration of the period between 1945 and 1973, the end of the widely held perception of eternal plenty posed a problem for advocates of environmentalism. Its values spoke to a different range of sentiments than did historic patterns of economic endeavor, differences that could be tolerated during periods of prosperity but that were easily challenged as the economy stagnated. The goals of environmentalism seemed to be class based and insensitive to more basic concerns, particularly among traditional, heavy-industry, blue-collar employees, where jobs became scarce and replacement employment came at tremendous decrease in remuneration.

At the same time, a new and far more ominous threat to the quality of life appeared in American society. As traditional industries slowed their growth, reorganized to the detriment of their blue-collar workforces, and prepared themselves for foreign competition, the toxic consequences of their operations came increasingly to the attention of the American public. Both hazardous and nuclear waste were recognized as threats not only to the American way of life but also to people's faith in their institutions. Such waste seemed omnipresent; industry, the military, and even Silicon Valley, the high-tech hope of the future of the American economy, produced a seemingly endless list of radioactive and nonradioactive hazardous by-products. When confronted with evidence of this reality, the public's belief in the system, already damaged by political scandal, took another beating. As the 1980s began, the growing distrust of politics in the post-Watergate era extended even deeper into American society.

Notes

1. Walter A. Rosenbaum, *The Politics of Environmental Concern* (New York: Praeger, 1973).
2. Hal K. Rothman, *The Greening of a Nation? Environmentalism in the U.S. since 1945* (New York: HarBrace Books, 1997).
3. Linda J. Lear, *Rachel Carson: Witness for Nature* (New York: Henry Holt, 1997).
4. Rachel Carson, *Silent Spring* (New York: Fawcett Crest, 1962).
5. Rothman, *The Greening of a Nation?*
6. Eric Goldman, *The Crucial Decade and After: America 1945–1960* (New York: Alfred A. Knopf, 1960).
7. Paul Ehrlich, *The Population Bomb* (New York: Ballantine Books, 1968).
8. Samuel P. Hays, *Beauty, Health, and Permanence: Environmental Politics in the U.S. 1955–1985* (Cambridge: Cambridge University Press, 1987).
9. Rothman, *The Greening of a Nation?*
10. Garrett Hardin, "The Tragedy of the Commons," *Science* 162 (1968): 1243–1248.
11. Ibid.
12. Lynton K. Caldwell, *Environment: A Challenge to Modern Society* (Garden City, N.Y.: Natural History Press, 1970), ix–xii.

Chapter 7

Social Science Research and Early Childhood Education

A Historical Analysis of Developments in Head Start, Kindergartens, and Day Care

KRISTEN D. NAWROTZKI, ANNA MILLS SMITH, AND MARIS VINOVSKIS

OVER THE SECOND HALF of the twentieth century the behavioral and social sciences played an important—although usually a secondary—role in helping policy-makers develop, implement, and evaluate early childhood education programs. While academic advisers to the federal, state, and local governments have rarely been able to prescribe the exact nature of a new program or how it should be run, these analysts have been frequently consulted as part of the decision-making process. Although the results of this collaboration between social scientists and government are not always satisfying or productive, over the last three decades those interactions have come to be expected, at least to some degree, in the creation of new government initiatives.[1]

One of the difficulties inherent in assessing the nature and impact of government social policies is that most are categorical and address only a small part of some larger problem or issue. As a result, policy-makers often fail to see the large picture or how the various policies might interact at the level of the client. Analysts frequently address specific initiatives in relative isolation from other related programs. Attempts to gain a more complete vision of these programs are further hampered by the physical separation of early childhood programs into multiple federal, state, and community departments.

In an effort to initiate a broader and more comprehensive analysis of early childhood education, we will examine briefly three related entities that are usually discussed separately: Head Start, kindergartens, and day care. Head Start is a federally funded, community-run compensatory preschool program designed to provide an enriched nursery school experience to disadvantaged children in the hopes of producing cognitive and social gains that will allow poor children to enter school with the same skills as their middle-class counterparts. Head Start has been traditionally viewed as an antipoverty program. Kindergarten, however, is an expected part of the public school system and is considered a normative experience for all children. Day care centers that provide care for children as young as two weeks are quickly becoming another common experience for children from all socioeconomic levels. Because it serves both children's physical needs and parents' employment needs, day care has occupied a nebulous third space, associated completely neither with welfare initiatives nor with educational programs.

Although the connections between Head Start, kindergarten, and day care are usually overlooked, it is our contention that these three programs merit an integrated analysis. Although Head Start, kindergarten, and day care are administratively and philosophically distinct, in the context of an individual child's life they are often intimately connected and together form the sum of many young children's first formal cognitive and social experiences. This is the primary reason for discussing these three programs together: all three address essentially the same population. The three programs are also linked historically in that they spring from the same premise that early childhood is a unique developmental stage that requires special attention and is easily compromised. In more recent times, all three were deeply affected by the Great Society initiatives of the mid–1960s, and all three have undergone significant rethinking over the past thirty years.

Finally, examining Head Start, kindergarten, and day care policies together provides a more comprehensive understanding of the interactions between social science research and early childhood policies than is possible when looking at each in isolation. Because of their essentially shared target population, the research conducted for one program usually has ripple effects on the other two. Similar debates over the nature of early childhood cognitive, social, and emotional development echo throughout all three. At the same time, each must continue to garner research that supports its unique contributions or risk elimination. Head Start, kindergartens, and day care all provide excellent examples of policies created by the coming together of political agendas and research findings. Because Americans have identified early childhood as a priority, each of these policies also represents the strong influence of social and cultural changes on programs designed for young children. While space and time do not permit an in-depth analysis of any of these activities, we will try to explore some areas that are particularly illustrative of this nexus between politics, social science research, and social environment.

Project Head Start

While most Americans think that early childhood education programs such as Head Start are recent inventions, in fact somewhat comparable efforts existed more than a 175 years ago. Infant schools, initially designed for disadvantaged children ages two to four in Great Britain, were imported into the United States in the mid–1820s. Infant schools were promoted by educational reformers on both sides of the Atlantic not only as a way of helping working mothers to care for their young children but also as a means of improving the education of the poor. Infant schools quickly became very popular throughout the United States as middle-class parents insisted that their own children receive the same benefits of early childhood education as their less fortunate counterparts. It is estimated by 1840 that nearly 40 percent of three-year-olds in Massachusetts were attending a public or private school.[2]

Early childhood education did not last long in antebellum America because in the early 1830s prominent medical authorities such as Amariah Brigham denounced early intellectual activity as threatening the healthy development of the young child. Brigham and his colleagues believed that premature intellectual stimulation diverted essential resources from a child's rapidly growing young brain, leading eventually to mental illness in youth and early adulthood. As a result, infant schools gradually lost much of their appeal and clientele; by the eve of the Civil War there were few children below the age of five in American schools.[3] When kindergartens were imported into the United States in the mid-nineteenth century, they emphasized the importance of play rather than teaching children the alphabet and how to read, and kindergartens tried to recruit slightly older children than the infant schools had served.[4]

Thus, while early childhood education programs like the infant schools had thrived in antebellum America, they quickly fell out of favor and were all but forgotten by nineteenth-century educators—as well as by most twentieth-century historians. And while there were efforts to provide some early childhood training through kindergartens and day nurseries in the early twentieth century, most children did not enroll on a regular basis.[5] Formal, full-time education for most children in the early decades of the twentieth century started with the first grade, usually at about age six.

During the 1930s and 1940s there was considerable debate over whether nursery schools could improve children's IQ scores. Some contended that improving a child's home and school environment would lead to higher IQ scores, while others maintained that heredity was the most important factor in establishing levels of intelligence.[6] By the 1950s the proponents of the heredity position seemed to be in ascendancy, but their conclusions now were challenged by influential scholars such as Benjamin Bloom and J. McVicker Hunt, who saw intelligence readily modifiable by a child's experiences—especially in the "critical period" of the first five years.[7] The belief that children's IQs could be improved also received reinforcement from several of the foundation-funded experimental early childhood programs in cities such as Baltimore, Nashville, New York, and Syracuse.[8]

The mid–1960s were a period of great change as President Lyndon Johnson and the newly elected Eighty-ninth Congress created the Great Society programs, which were intended to help disadvantaged Americans escape from the "cycle of poverty" that had trapped them. While the political and social climate of the time was conducive to large-scale federal domestic innovations in general, there were also several other developments that made early childhood programs such as Head Start particularly attractive to policy-makers.[9]

The work of developmental psychologists such as Bloom and Hunt, as well as the foundation-funded community childhood programs, provided an important framework for Head Start, but the major impetus for that initiative came from policy-makers such as Sargent Shriver, director of the Office of Economic Opportunity (OEO), who were looking for politically attractive ways for the federal government to help at-risk children escape poverty. Ignoring the recommendations of several of his key academic advisers to pilot the program first, in 1965 Sargent Shriver initiated a large-scale, eight-week summer program for at-risk children. Head Start was intended to be a broad, comprehensive federal program that provided a variety of services to children (including health services) and emphasized the involvement of parents. While many of the academic advisers downplayed the expectations for the educational component of Head Start, Shriver told Congress that the new initiative might be able to raise children's IQs by eight to ten points.[10]

From the very beginning Head Start was one of the most popular Great Society programs, with its emphasis on helping young children and the willingness of the federal government to cover almost the entire cost of the enterprise (unlike the situation in K–12 education where the new federal initiatives provided less than 10 percent of the overall costs). However, a major evaluation of Head Start by the Westinghouse Learning Corporation and Ohio University argued that the IQ gains from the program were only temporary and faded by the time children entered the regular schools. The authors did go on to point out that there were important noncognitive and nonaffective benefits from the program.[11]

While the Westinghouse report's perceived negative conclusions were strongly challenged by other scholars, policy-makers began to question Head

Start's efficacy and shifted its emphasis from a summer program to a year-round endeavor.[12] Parents involved in the Head Start program continued to support the initiative notwithstanding the Westinghouse critique and successfully mobilized political support on its behalf. Thus Head Start, unlike the nineteenth-century infant schools, withstood a major challenge to its survival; but policy-makers remained more reserved about its efficacy, and funding for the program in real dollars stagnated in the 1970s and 1980s.[13]

Even as Head Start was created in the summer of 1965, there were questions about the transition of the enrolled children into regular classrooms. These concerns were reinforced by the private doubts of Edward Zigler, a prominent early childhood psychologist, about the ability of Head Start to dramatically improve children's IQs. In addition, a small but influential study by Max Wolf and Annie Stein of a sample of children who had attended Head Start in the summer of 1965 found that the IQ gains faded quickly. Thus, almost from the beginning, several analysts expressed doubts about the ability of the summer Head Start program to improve significantly the intelligence of at-risk children as well as to help them make a smooth transition into the regular schools.[14]

Responding to the concerns about the apparent limitations of Head Start, as well as eager to improve K–12 education, President Johnson proposed in 1967 that Congress create a "follow through" program to help children make the transition from Head Start to the early years of formal schooling. Congress responded positively and legislated the Follow Through Program in the 1967 amendments to the Economic Opportunity Act of 1964. Follow Through was to be a large-scale education service program with 200,000 children anticipated for enrollment in the 1968–1969 school year at a cost of about $120 million.[15]

The financial pressures of the Vietnam War as well as the growing skepticism about the Office of Economic Opportunity (OEO) limited the FY1969 new funding for Follow Through to only $12 million. Given the unexpectedly limited funding, Follow Through was converted to an experimental program in principle—though in practice many of its advocates continued to treat it as a small-scale service program that they hoped would expand in the near future.

In its new incarnation Follow Through was designed as a planned variation experiment—local programs would be varied along some dimension such as curriculum, student-teacher ratios, teacher characteristics, or type of parental involvement in order to see what worked best in helping disadvantaged children make the transition into regular schools. Unfortunately, the confusion between an experimental program and a service one was never fully resolved, and the evaluations of Follow Through were not as scientifically sound as had been anticipated. Moreover, the substantive findings generally were disappointing, as Follow Through was unable to help at-risk children close much of their achievement gap relative to their more advantaged compatriots. Nor did most of the individual models tested prove to be particularly effective—especially when the same model was tried in different settings.[16]

Given the disappointing results as well as the questions that had been raised about the quality of the national evaluations of Follow Through, Republican and Democratic administrations both made repeated efforts after the mid–1970s to terminate the program, but strong congressional support for Follow Through from legislators whose districts received these funds maintained a much-reduced Follow Through program into the early 1990s. Altogether, Follow Through was one of the more expensive early childhood education experiments—costing more that $1.5 billion (in constant 1982–1984 dollars) from 1967 to 1994. Despite the large sums of monies spent, few educators or policy-makers today are even aware of the Follow Through program. Nor have the results from that expenditure provided us with much guidance on how to help children in Head Start make a more effective transition into regular classrooms—an issue that today still waits to be addressed further.[17]

During the 1980s policy-makers' interest in Head Start revived as an increasing proportion of mothers entered the labor force and as the results from other existing federal compensatory education programs, such as Title I of the Elementary and Secondary Education Act (ESEA) of 1965, continued to be disappointing. Early childhood education also benefited from the publication of several studies of model programs that claimed to demonstrate the long-term success of these efforts.

One particularly interesting study was the analysis of the highly controversial but very influential Perry Preschool program in Ypsilanti, Michigan. This small-scale, high-quality program was started in the mid–1960s and was one of the very few early childhood evaluations to employ a randomized control group. Although the results from the Perry Preschool study also indicated an initial fading of the cognitive gains, over time the program reduced the levels of juvenile delinquency and teenage pregnancy among the participants. Reminiscent of the claims on behalf of the nineteenth-century infant schools, early childhood advocates emphasized the cost-effectiveness of this intervention in preventing subsequent problems, and its developers called for its expansion throughout the nation.[18]

While most observers have praised the effectiveness of the Perry Preschool program, some have questioned its broader applicability. They point out that the high-quality care provided in the Perry Preschool program is not similar to that present in most Head Start programs today. They also question the statistical design of the evaluation because there were some significant differences between the control and the intervention groups. Others have raised questions about the ways in which the cost-benefit analyses have been conducted and reported.[19] Perhaps most troubling is the recent, belated revelation that there were significant gender differences in outcomes. While the program seems to have been quite beneficial for the girls, it was much less so for the boys (often there were no long-term differences for the males participating in the program).[20] Unfortunately, this important finding about the gender differences in the effectiveness of the Perry Preschool project has not been widely publicized, so that most of the public and policy-makers who cite the study are unaware of this potentially highly significant differential.[21]

Public support for Head Start today is probably at an all-time high, and many policy-makers now vie with each other to see who is the strongest supporter of the program. Both Presidents George H. W. Bush and Bill Clinton strongly endorsed Head Start, and funding for that program increased accordingly.[22]

Yet some of the recent assessments of Head Start continue to raise questions about its long-term effectiveness. On the one hand, Ron Haskins, a developmental psychologist and at the time a congressional staff member, reviewed the existing studies in 1989 and found that while there were intellectual and socioemotional gains during the first year, most of them faded over time.[23] On the other hand, Stephen Barnett, a prolific scholar who had coauthored some of the Perry Preschool analyses, is more optimistic about these programs in general—though he readily acknowledges that there were significant differences between the high-quality interventions and the typical Head Start programs.[24] Moreover, other scholars point out that even in high-quality programs like the Perry Preschool program, the intervention was not sufficient to close the large academic gap between the advantaged and disadvantaged students.[25]

One recent suggestion for improving Head Start is to enhance its education components in order to assist at-risk students in learning to read. During the 2000 presidential election, candidate George W. Bush called for transferring Head Start into the U.S. Department of Education and emphasized fostering literacy skills in early childhood education programs.[26] Once in office and having encountered considerable opposition to moving Head Start, the Bush administration backed away from its pledge to transfer the program.[27] But President Bush has reiterated his belief that preschool education in general and Head Start in particular should place more emphasis on preparing young children to begin to learn how to read.[28]

The major federal compensatory education program, the ESEA, was reauthorized as the No Child Left Behind Act of 2001. The legislation included authorization for $900 million for Reading First in order to help states create "scientific, research-based" reading programs for students in the K–3 grades; the bill also authorized $75 million for the new Early Reading First Program to help three-to-five-year-olds learn to read.[29] Moreover, President Bush called for training all Head Start teachers to teach early literacy and for ensuring that Head Start centers meet the learning standards established in the program's reauthorization in 1998.[30]

While the recent calls for introducing more early reading instruction have received many favorable comments as well as substantial federal support, they are also encountering a growing backlash among some policy-makers and researchers. Early childhood education advocates criticize the Bush administration's proposed 1.9 percent increase for FY2003 Head Start funding as inadequate.[31] And some educators and researchers question the "scientific and research-based" evidence used to justify the new federal reading initiatives.[32] For example, University of Florida professor Richard Allington, past president of the National Reading Conference, complains that "the scientific evidence we do have about teaching and learning to read now is being selectively reviewed, distorted, and misrepresented by the very agents and agencies who should give us reliable reports of what the research says. I am worried because ideology is trumping evidence at the moment and teaching and learning to read will be [*sic*] both be worse for it."[33]

Although the Clinton administration had argued that there was no need for additional evaluations of the effectiveness of Head Start and similar programs since earlier studies had established their effectiveness, the Government Accounting Office's (GAO) review of those analyses reaches an opposite conclusion. The GAO called for additional, scientifically sound studies of Head Start, and the U.S. Senate's Budget Committee's Task Force on Education agreed with its assessment.[34]

The Bush administration is expanding research on early childhood education. Grover J. "Russ" Whitehurst, the new assistant secretary of the Office for Educational Research and Improvement (OERI), at the annual meeting of the National Association for the Education of Young Children (NAEYC), criticized the early childhood education field for paying too little attention to evaluation.[35] The Bush administration proposed a five-year $50 million analysis of ways to improve children's development and foster school readiness.[36] Thus, while there continues to be some real differences among scholars and policy-makers about the effectiveness of different approaches to early childhood education, there are a number of useful studies under way that will be quite helpful when they are completed.[37]

Kindergartens

Many of the social scientific techniques and assumptions that underlay the development of Head Start since the 1960s also influenced the kindergarten but did so more slowly and much less directly. Part of local public school systems, kindergartens are locally funded and locally administered programs whose design and quality vary widely. Compared to Head Start, the kindergarten as an institution had much less clearly articulated goals to begin with and garnered much less attention and funding on the federal level.[38] Unlike the new Great Society programs, the kindergarten's existence and expansion were not under debate in the 1960s and 1970s. On the contrary, the value of the kindergarten was largely seen by parents and policy-makers as "given" despite the fact that by 1964, less than 30 percent of U.S. public school districts offered kindergarten education.[39] As a result, research on the kindergarten itself has been relatively sparse; more often, the results of other early childhood studies (including those on Head Start) have been extrapolated to describe the probable experiences of kindergarten children.[40] State and federal policy attention, too, has been more focused on prekindergarten than on kindergarten programs.

Only in the last twenty years have social scientists focused specifically on the kindergarten, raising questions about its effectiveness and indeed about its very purpose as part of a child's educational and social development. By the end of the 1970s increased socioeconomic heterogeneity of children attending kindergarten made Great Society concerns more relevant to the kindergarten, situating the kindergarten itself at the center of early childhood and primary grades education reform. Beginning in the 1980s, studies on the kindergarten focused on two major issues: quantity and quality, both of which are determined at the state and local levels.

When kindergartens were added to public schools in the early twentieth century, kindergarten teachers taught two half-day sessions each day in order to accommodate large numbers of children.[41] For the most part, public school kindergartens continued to run half-day programs through midcentury, a trend furthered by baby-boom population pressures. Since the 1970s, however, increasing numbers of school districts have begun to provide kindergarten programs lasting an entire school-day, approximately six hours.

The trend toward full-day kindergartens really took off in the 1980s as parents looked to the schools for child care and as schools sought to minimize the transportation costs and scheduling burdens associated with traditional half-day kindergartens.[42] By 1985 surveys of kindergarten provision showed four main types of kindergartens nationwide. Of all kindergarten programs, 67 percent were half-day and 22 percent were full-day programs with each child attending daily, while 7 percent were full-day programs which children attended on alternating days. Presumably, the remaining 4 percent of kindergartens were organized on still other schedules.[43]

It soon became obvious, however, that longer hours did not translate into better kindergarten programs.[44] In 1983 social scientists with the Consortium for Longitudinal Studies found that full-day kindergartens especially aided the later academic performance of low-income and non-English-speaking children, but the generalizability of such claims was admittedly problematic.[45] Full-day kindergartens were found to have some long-term effects on some children's subsequent school performance, but in the absence of enhanced curricula and pedagogy, such gains were nonexistent. Other longitudinal research, such as that emerging from the Beginning School Study (BSS), also claimed that the number of hours a child attended kindergarten each day positively affected cognitive outcomes for African American children early in the first grade. However, BSS results also indicated that increased kindergarten attendance did not significantly improve African American children's socioemotional development, nor did it influence white children either cognitively or affectively. Moreover, the cognitive effects that were experienced by African American children diminished significantly by the end of the first grade.[46] This led researchers to believe that, as with Head Start, kindergarten attendance seemed to have initially strong effects on some children, but those effects were ultimately short-lived.

Still, the full-day kindergarten managed to gain widespread public and political support in the 1980s and 1990s, even as the educational research community hotly debated its universal usefulness. While parents, teachers, and policy-makers in some parts of America engaged in debate about the costs and benefits of full-day versus half-day kindergartens, people in other states and localities were still struggling to get even half-day kindergarten education to be offered as part of the public schools.[47] The trend toward full-day public kindergarten programs was in accordance with other moves (also bolstered by ambiguous research results) to add hours to students' schooling by lengthening school years or moving to year-round schedules. An increase in public and private full-day prekindergarten provision for three- and four-year-olds was also part of this trend. These changes in kindergarten scheduling and in other aspects of schooling heightened long-term debates regarding the kindergarten's purpose in educating young children.[48] And, as research on full-day kindergartens demonstrated, quantity of schooling meant little unless quality could be assured.

From its nineteenth-century roots, the American kindergarten was generally accepted as a reaction against the academic emphases of its predecessor, the infant school. Unlike the early-nineteenth-century infant school, the mid-nineteenth-century kindergarten eschewed literacy training and focused instead on socializing young children to each other, to the school, and to American society as a whole. By the late nineteenth century, then, American parents and newly arrived "child experts" seemed to have agreed that young children's training should focus on socialization and that literacy should come later on. The tension between academic and socialization emphases flared again, however, as the kindergarten sought its place in local programs of public education at the start of the twentieth century. Wrangling about the purposes of kindergarten education continued at various levels throughout the century.

From the 1970s, advances in the use of social science for program planning and assessment encouraged new efforts toward striking the right balance between literacy and socialization foci in the kindergarten. To this, a new goal was added in the 1990s, as educators and researchers sought ways to smooth the transition from preschool, through kindergarten, and into primary grades in order to maintain apparent gains made at each level. Those concerned with high-quality kindergarten curricula and pedagogy generally split into two main camps: those focusing on literacy or early academic training and those emphasizing socialization.[49] From about the 1960s, many parents took to the idea that the old kindergarten work (that is, socialization) was being accomplished in the modern preschool, making the kindergarten the place to start academic training. Increased parental voice in school decision-making and comparisons with the Japanese educational system in educational journals and the popular press also contributed to this sweeping change.[50]

The long-term trend of "beefing up" the cognitive aspect of kindergarten curriculum with academic training meshed with the excellence reform movement in the 1980s.[51] Triggered in part by the U.S. Department of Education's 1983 report, *A Nation at Risk*, the movement saw a back-to-basics educational approach as a means to ensure U.S. economic dominance in world markets.[52] President George H. W. Bush and the nation's governors eventually responded to the suggestions of *A Nation at Risk* by formulating national goals for education, including that at the preschool level.[53] The Clinton administration continued this effort, garnering bipartisan support of what became the Goals 2000 Act of 1994. Goal One of the act ambitiously promised that "by the year 2000, all children in America will start school ready to learn," thereby improving the efficiency and efficacy of education at all levels.[54]

While debates continued over what (if anything) made early schooling both efficient and effective, the push for raised academic standards in early education did not go unchallenged. In the late 1980s early childhood professionals spoke out about the dangers of overly academic approaches and the misuse of testing, retention, and grouping practices in early education.[55] Policymakers took on board the claims of social scientists who said that the growing emphasis on school effectiveness, accompanied by high-stakes testing of kindergarten readiness and early grades performance, was detrimental to the youngest students.[56] Academic frustration in the kindergarten led to academic failure later on, and it seemed to many that American children were "losing their childhoods" to academic pressures.

The National Association for the Education of Young Children (NAEYC) led the charge against academic pressures for young children in the 1980s, identifying developmentally appropriate and inappropriate practices for young children through age eight.[57] NAEYC guidelines precluded the use of didactic, teacher-centered pedagogy for three-to-five-year-olds, providing instead a "print-rich" environment without actually training them to read per se.[58] Although a majority of educators and researchers found the NAEYC's guidelines to be both theoretically reasonable and pragmatically sound, a few critics found them to be "social constructed, contextbound, and insensitive to cultural and individual differences in development."[59] In the absence of alternatives, the NAEYC's standards remained popular through the 1990s even as they were revised and updated.

Still, new social science–based efforts toward defining high-quality early childhood education meant new possibilities for resolving debates about whether kindergarten children should learn to read. In the late 1990s the National Academy of Sciences (NAS) Committee on the Prevention of Reading Difficulties in Young Children built upon the NAEYC's earlier work to reiterate warnings against fast-paced early academic training. It identified the kindergarten as extremely important to the development of reading skills but did not see it as a "hothouse" in which to "force" children's reading abilities to bloom early.[60] Although the NAEYC and NAS reports offered concrete strategies to kindergarten teachers and school administrators, surveys indicated that a wide gap remained between their recommendations and what was really happening in classrooms across the country.[61]

Later, the aforementioned No Child Left Behind Act of 2001 promised to narrow this gap by translating scientifically based reading research directly into instructions for classroom practice. Of course, the act's Reading First and

Early Reading First programs, by definition, focused on the achievement of literacy rather than the other cognitive and affective early education goals that advocates of developmentally appropriate practices tried to protect. In the 1980s the clear identification of developmentally appropriate practice was intended to speak to the quality and quantity concerns raised about kindergartens. In the 1990s the issue of quality continued at the forefront of research related to the kindergarten, while studies (such as the National Transition Study) looked at whether Head Start or kindergarten gains could be extended into the next levels of schooling.[62] Although the kindergarten year was at the center of these studies, none of them looked at what kindergartens were actually like.

To help fill this gap, the U.S. Department of Education's National Center for Education Statistics began the Early Childhood Longitudinal Study Kindergarten Class of 1998–99 (ECLS-K) study. This study was the first national longitudinal study to track children as individuals from the time they enter kindergarten.[63] Jerry West, the study's director, noted at the study's outset that although "[m]ost children are now going to kindergarten, . . . we have relatively little information on a national level on what that experience is like for them."[64] Researchers and educators alike hoped that the study would finally illuminate children's "experiences while in kindergarten, and . . . how those experiences relate to later success in school."[65] With this information, then, steps could be made toward a greater unification of planning and assessment of kindergarten curricula and pedagogy on a large scale, combining analysis and evaluation techniques of social science, the practical aspects of pedagogy, and the sweeping reach of policy. Finally, they hoped that data gathered from the 22,000 ECLS-K participants would help researchers to overcome the problems of implementation and evaluation caused by the decentralized nature of school administration.

As of late 2002 the ECLS-K project had yielded three major reports on kindergarten children in America.[66] In addition to providing important descriptive data on who goes to kindergarten, where, and for how long, these reports summarized differences in cognitive growth, skills attainment, and affective development in children before, during, and after their kindergarten experiences. Unsurprisingly, children come to kindergarten with different amounts of knowledge and skill, depending upon developmental factors such as their own age and maturity as well as environmental factors related to the nature of their home lives and their child care histories.[67] Though, as expected, individual children gain particular types of literacy and mathematics skills at different rates, the ECLS-K's 2001 report found a general consistency in the extent of children's overall gains in cognitive and social knowledge and skill development by the end of the kindergarten year. According to the ECLS-K data, there appeared to be little difference in the gains made by children enrolled in public versus private kindergartens or attending full-day as opposed to half-day programs.

The ultimate goal of this kindergarten-specific research is to allow for the development of high-quality, developmentally appropriate kindergarten curriculum models flexible enough to suit local needs yet standardized enough to allow for programs to be assessed accurately. One thing we have learned from increasingly sophisticated research is how to rephrase the major questions; that is, the research framework should not be one of quality versus quantity but rather how to maximize both to serve the needs of children, parents, and schools.

The most recent efforts at kindergarten-specific research recognize the unique place the kindergarten occupies in American education and its relationship to other aspects of early childhood and later development. This research challenges the popular assumption of the kindergarten as a "given," as a necessarily coherent program of early childhood education or as a unifying bridge between home and school. After all, the curricula and pedagogy of kindergartens vary as widely as their clientele, including children who have attended Head Start or other preschool programs, children who have been in public or private nonfamilial day care settings, and those who have been exclusively in the care of a parent or family member for their first four or five years of life. In addition, kindergarten teachers have to react to changes in federal and state curriculum policy while responding to research-based changes in the state of the pedagogical art.[68] Given the federal government's attention to Head Start and other early childhood programs, it seems only natural that it should support and respond to research and development on the kindergarten as well.

Day Care Programs

Publicly funded child care services over the last century have been limited, inconsistently funded, and philosophically incoherent. The history of these services can best be understood by examining a few characteristics of day care policy development that have remained persistent throughout the twentieth century. Day care services provide custodial care to children who for various reasons cannot be cared for at home or school. While many child care services also have educational and developmental goals, most exist to enable parents (usually mothers) to work outside of the home. Day care, therefore, is not primarily an educational initiative, a child welfare service, or an employment incentive but rather contains aspects of all three. Because day care policy must address such fraught topics as public welfare spending, family structure, and appropriate gender roles, it is also particularly subject to the cultural, social, and scientific influences of its historical context.

It is frequently difficult to identify past day care initiatives, let alone to coherently discuss their connection to contemporary social science research. This difficulty is further exacerbated by the bureaucratic and ideological tensions that have emerged as a result of the multifaceted nature of day care services. For example, there has been an ongoing tension between day care programs designed to protect the children of working parents and early childhood educational programs, such as Head Start, intended to enrich children's cognitive, emotional, or social development. A similar tension exists between day care services and larger public welfare programs. An examination of how social science research influenced child care policy decisions must include multiple, competing federal agencies and social science disciplines.[69]

World War II saw the first widespread publicly funded day care services. Between 1942 and 1946 the United States government funded 3,102 day care centers serving 600,000 children in forty-seven states. In the context of a pressing need for female labor and intense fears about masses of unsupervised children, the call for day care policy briefly overcame years of resistance and stigma. However, throughout the duration of the war, day care policy remained hotly contested and mired in philosophical and political debates. Education, child welfare, and public works experts all sought to claim wartime child care as their domain. However, as the war ended, no federal-level agency claimed day care as an ongoing priority. Private foundations and civic organizations similarly neglected day care in their postwar policy agendas. Professional social workers, educators, and child psychologists who had cautiously supported day care centers as a necessary evil during the war quickly reversed their opinions.[70] The assumption that the highest ideal of American family life was a well-paid father and a home-based mother had survived the war intact and was at the core of postwar domestic policy-making. Furthermore, a profusion of new social science research and writing provided an intellectual foundation for these beliefs.

Although the working mother was a very real phenomenon, the dominant discourse after the war labeled her as aberrant. Books, magazines, movies, and television all glorified the female homemaker. With the burgeoning baby boom and new thinkers such as Dr. Benjamin Spock, child rearing became perceived as a sacred craft. New changes in psychotherapy focused attention on mothers as the guardians of their children's mental health and heterosexuality. There was simply no room amid this resurgence of the cult of motherhood for a day care policy that might have challenged the supremacy of maternal child rearing.[71] Moreover, as John Bowlby's studies on the effects of maternal deprivation on English war orphans began to circulate, nonmaternal child care, especially institutional care, was effectively eliminated as a possible policy response to the growing number of working mothers.[72]

Day care did not reemerge as a federal issue until the early 1960s, amid growing concerns over increasing welfare roles and entrenched poverty. Social workers and child welfare professionals had long recognized the contradictory relationship between day care and Aid to Dependent Children legislation—the former an institution that allows mothers to work and the latter an institution that expects them not to work. In the 1960s day care, instead of being a threat to the renamed Aid to Families with Dependent Children (AFDC) program, came to be viewed as part of its solution. Through the Social Security Act, federal money was made available to subsidize the care of young children receiving AFDC so that their mothers could work or attend job training. Day care was also classified as a necessary welfare service for abused children or children with disabilities and as a result became part of the Great Society measures aimed at eliminating poverty. At the same time, however, researchers continued to rely on Bowlby's theories of separation and attachment to condemn nonmaternal care for infants, for the children of working poor parents, and for children whose mothers were not forced to work outside the home for economic reasons.[73]

Although day care services did not directly benefit from the popularity of Head Start, the program and the research that had inspired it created a greater overall interest in early childhood. Out of this new enthusiasm came a reshuffling of agencies dealing with children at the federal level, including the creation of a new Office of Child Development (OCD). Ed Zigler of the OCD acknowledged the low success rate of child care policy initiatives and pointed to the dearth of expertise among day care supporters. Unlike the movements for Head Start, the kindergarten, and welfare reform, the day care field was populated by low-paid, nonprofessional staff who had little research data and few resources at their command.[74]

In 1971 Senator Walter Mondale (Democrat from Minnesota) and Representative John Brademas (Democrat from Indiana) sponsored the Comprehensive Child Development Act, which would have increased funding for (and the availability of) both early childhood education programs (such as Head Start) and protective day care services for the children of working mothers. The proposed legislation would have for the first time explicitly integrated the goals of both types of early childhood programs. While both chambers of Congress approved the legislation, it was vetoed by President Nixon, who claimed that for all of its good intentions, it was "overshadowed by the fiscal irresponsibility, administrative unworkability, and family weakening implications of the system it envisions." Although the marked increase in the numbers of working mothers of young children continued unabated, persistent discomfort with the possible implications of nonmaternal, institution-based care remained an obstacle.[75]

In the years that followed, several subsequent pieces of legislation with similar aims failed to make it through Congress. There was, however, a significant increase in the amount of private research on the effects of nonmaternal care on young children. Multiple, competing studies produced contradictory results about the safety and advisability of formal child care settings. Results varied based on the age group studied and the criteria used to determine children's mental health, always reflecting deep societal ambivalence about employment for mothers of young children.[76]

The 1980s saw drastic cutbacks in the amount of federal money allocated to subsidize day care for families receiving AFDC and for other very low income families. Funding for Title XX of the Social Security Act, which provided the money for most of these subsidies, was inconsistent throughout the 1980s. The emphasis remained on day care as a custodial service for poor families, with little attempt to incorporate the new knowledge of early childhood development that had spurred the creation of Head Start and other enrichment programs. In 1990 a new federal child care bill was introduced to increase AFDC subsidies, provide two large block grants to fund more centers for low-income and working families, and to expand child care income tax credits for these families.[77]

The Personal Responsibility and Work Opportunity Reconciliation Act of 1996 (PRWORA), the welfare reform law passed during the first Clinton administration, significantly reshaped federal child care policy. As AFDC was abolished, all previous sources of federal child care funding were replaced with the single, newly revised Child Care and Development Block Grant (CCDBG). The CCDBG provides states with two funding streams, a capped entitlement program requiring matching funds and a smaller discretionary program.[78] The Child Care Bureau, which had been established within the Administration for Children, Youth and Families in 1995 as an attempt to provide a focus for federal child care policies, now took over the much larger responsibility of distributing and regulating all of the funds provided under the CCDBG. While there was much partisan rancor over both the larger ramifications of the PRWORA and the specific changes made in child care policy, this law firmly institutionalized child care within the federal government for the first time.[79]

By the late 1990s there was an unprecedented wealth of new research available on the effects of nonmaternal child care. These new studies came out of increased public interest in early childhood development, the continuing rise in the number of employed mothers with young children, and the establishment of an advocacy community specifically devoted to the issue of quality child care. However, interpretations of the new data were highly politicized, and the results of conflicting studies were pitted against each other in the political arena.

One of the most prominent of these studies was a longitudinal study sponsored by the National Institute of Child Health and Human Development, which determined that it was the quality of day care that best predicted the effects of day care on child development. Child care advocacy groups, such as the Children's Defense Fund, used these results to call for increased federal regulation and training of day care providers to improve the quality of care. Conservative advocacy groups contended that increased federal regulation would only further curtail individual families' latitude to determine the best kind of care for their children and would encourage nonparental care as an acceptable alternative for all families rather than an unfortunate necessity for poor families.[80]

Current child care policy reflects the legacy of day care's history of stigma, incoherence, and ambivalence. While it has now become politically untenable for either party to ignore the importance of child care in the lives of American families, disputes over the role of the federal government in child care remain contentious. Recently, these partisan and philosophical divisions have been further strained by the George W. Bush administration's support for funding faith-based initiatives, including child care centers.

Although social science research has provided many new perspectives and new information on the role of day care services in children's lives, little of this research has been reflected in policy. The Child Care Bureau can collect resources on what constitutes quality care but can only enforce very specific health and safety regulations.[81] Both the Clinton and George W. Bush administrations have advocated better integrating child care services with other early child education programs to provide a seamless continuum of early childhood experiences, but federal child care policy remains focused on low-income families.

The expansion of public day care has been consistently driven by immediate economic and political needs—such as the need for wartime labor or the call to limit welfare rolls. Similarly, although research has been used to both praise and condemn day care, it seems that the most powerful forces influencing day care remain public opinion and private activism, not the social and behavioral sciences.

Conclusion

Head Start, kindergarten, and day care operate at different levels of policy (federal, state, local, private, and public) and have been influenced by social science research to varying degrees over the last thirty-five years. The federal support garnered by Head Start and related programs since the 1960s has spurred an unprecedented amount of research and development, but such work often has been weak in terms of study quality and reliability of results. One of the best studies may be that on the Perry Preschool High/Scope project, but even that analysis has shown its weaknesses and a lack of generalizable findings. This is especially problematic because policy-makers and others have relied on that and numerous other ambiguous compensatory education studies to prescribe changes in day care and other educational and early childhood programs and policies. Only recently has the kindergarten been systematically included in research about early childhood education and transitions, and day care is still often neglected as a component in the intellectual (as opposed to socioemotional) development of young children.

More important than the differences in Head Start, kindergartens, and day care, however, is what they have in common. Concerns about early childhood education and care have existed long before the mid-twentieth century, but only since that time have they come to center stage in both policy and research. Head Start, kindergartens, and day care programs all serve the same or similar clientele. All three have been shaped to some degree by psychological models of brain development that emphasize the importance of early intellectual stimulus and emotional growth. These models stress the unique nature and malleability of young children's minds and have been used alternately to support and decry early academic work and literacy training as far back as the nineteenth century. Similarly, developmental psychological models have been used alternately to praise and condemn publicly funded day care programs, which have always been torn between public opinion, economic exigencies, and personal needs.

At the same time, however, the life-course perspective has been overlooked in much of early childhood research and policy. Instead of emphasizing early intervention at all costs, the life-course perspective suggests that lifetime effects are dependent on a variety of factors and experiences over the long term. Inklings of the life-course perspective may be seen in policy efforts such as Follow Through and recent research on early childhood transitions, but the conflict continues between that long-term emphasis and the developmental models stressing early intervention as the only solution. Some of the more effective intervention programs, such as Robert Slavin's "Success for

All," have explicitly raised questions about the efficacy and centrality of early childhood programs such as Head Start.

The Bush administration and the 107th Congress now are trying to expand the early education components in Head Start, kindergartens, and day care programs. These initiatives are generating considerable debate over the wisdom of stressing early literacy skills as well as questions about how best to achieve this goal. In the process, however, there is more attention to the interconnections among Head Start, kindergartens, and day care programs. Whether this new emphasis will lead to more integration of the three programs as well as improvements in the academic skills and well-being of young children remains to be seen.

Despite disagreement about the best way to care for and educate young children, the fact remains that early childhood has become both a research and policy priority in recent decades. Indeed, early childhood education, care, and research have become industries unto themselves, with something of an unstable relationship existing between them. The prioritization of some aspects of early childhood may be seen as a response to social changes, including changing ideas about child development, international economic competition, gendered shifts in the domestic labor market, and increasing awareness of structural and cyclical poverty. Still, while Head Start and related programs retain federal support and as kindergartens are gaining ground as a subject of research and public attention, day care has only recently become a significant subject of research and policy debates.

Over the last thirty-five years, then, we have seen a clear dependence on social scientific research to legitimate new policies and to justify dollars already promised or spent. But why do we give so much money and attention to early childhood to begin with? By and large, researchers, policy-makers, teachers, and parents have come to believe that early childhood is a unique phase of life and that early intellectual and emotional stimulation and protection can improve the quality of life and potential for young children. Thus to some degree American dedication to early childhood research and policy has been a reflection of appeals to social justice by attempting to reduce social inequalities through compensation for early disadvantage. Although research results and the work of early childhood programs in operation do not necessarily bear that out, we continue to operate on that assumption.

Notes

1. On the growing literature on the social and behavioral sciences in the formulation of government domestic policies, see David L. Featherman and Maris A.

Vinovskis, "Growth and Use of Social and Behavioral Science in the Federal Government Since World War II," in *The Social Sciences and Policymaking*, ed. David L. Featherman and Maris A. Vinovskis (Ann Arbor: University of Michigan Press, 2001), 40–82; Gene M. Lyons, *The Uneasy Partnership: Social Science and the Federal Government in the Twentieth Century* (New York: Russell Sage Foundation, 1969); Richard P. Nathan, *Social Science in Government: Uses and Misuses* (New York: Basic Books, 1988); Walter Williams, *Mismanaging America: The Rise of the Anti-Analytic Presidency* (Lawrence: University Press of Kansas, 1990); Robert C. Wood, *Whatever Possessed the President? Academic Experts and Presidential Policy, 1960–1988* (Amherst: University of Massachusetts Press, 1993).

2. John W. Jenkins, "Infant Schools and the Development of Public Primary Schools in Selected American Cities before the Civil War" (Ph.D. diss., University of Wisconsin, 1978); Dean May and Maris A. Vinovskis, "A Ray of Millennial Light: Early Education and Social Reform in the Infant School Movement in Massachusetts, 1826–1840," in *Family and Kin in American Urban Communities, 1800–1940*, ed. Tamara K. Hareven (New York: Watts, 1977), 62–99.
3. Carl F. Kaestle and Maris A. Vinovskis, *Education and Social Change in Nineteenth-Century Massachusetts* (Cambridge: Cambridge University Press, 1980).
4. Barbara Beatty, *Preschool Education in America: The Culture of Young Children from the Colonial Era to the Present* (New Haven, Conn.: Yale University Press, 1995); Sonya Michel, *Children's Interests/Mothers Rights: The Shaping of America's Child Care Policy* (New Haven, Conn.: Yale University Press, 1999); Carolyn Winterer, "Avoiding a 'Hothouse System of Education': Kindergartens and the Problem of Insanity, 1860–1890," *History of Education Quarterly* 32 (1992): 289–314.
5. Kristen Dombkowski, "The Kindergarten Movements in Victorian and Edwardian England and America" (M. Phil. Thesis, Oxford University, 1997).
6. Hamilton Cravens, *Before Head Start: The Iowa Station and America's Children* (Chapel Hill: University of North Carolina Press, 1993); Victoria L. Getis and Maris A. Vinovskis, "History of Child Care in the United States before 1850," in *Child Care in Context: Cross-Cultural Perspectives*, ed. Michael E. Lamb, Kathleen J. Sternberg, Carl-Phillip Hwang, and Anders G. Broberg (Hillsdale, N.J.: Lawrence Erlbaum, 1992), 185–206.
7. Benjamin S. Bloom, *Stability and Change in Human Characteristics* (New York: John Wiley, 1964); J. McVicker Hunt, *Intelligence and Experience* (New York: Ronald Press, 1961).
8. Edward Zigler and Karen Anderson, "An Idea Whose Time Had Come: The Intellectual and Political Climate for Head Start," in *Project Head Start: A Legacy of the War on Poverty*, ed. Edward Zigler and Jeanette Valentine (New York: Free Press, 1979), 3–19.
9. On the development of the Great Society programs, see Stephen K. Bailey and Edith K. Mosher, *ESEA: The Office of Education Administers a Law* (Syracuse, N.Y.: Syracuse University Press, 1968); Irving Bernstein, *Guns or Butter: The Presidency of Lyndon Johnson* (New York: Oxford University Press, 1996); Marshall Kaplan and Peggy Cuciti, eds., *The Great Society and Its Legacy: Twenty Years of U.S. Social Policy* (Durham, N.C.: Duke University Press, 1986); Sar A. Levitan and Robert Taggart, *The Promise of Greatness: The Social Programs of the Last Decade and Their Major Achievements* (Cambridge: Harvard University Press, 1976).
10. On the creation of the Head Start program, see Maris A. Vinovskis, "Early Childhood Education: Then and Now," *Daedalus* 122 (1993): 151–175; Sheldon H. White and Deborah W. Phillips, "Designing Head Start: Roles Played by Developmental Psychologists," in Featherman and Vinovskis, *The Social Sciences and*

Policymaking, 83–118; Zigler and Valentine, *Project Head Start*; Edward Zigler and Susan Muenchow, *Head Start: The Inside Story of America's Most Successful Educational Experiment* (New York: Basic Books, 1992).
11. Westinghouse Learning Corporation, "The Impact of Head Start: An Evaluation of the Effects of Head Start on Children's Cognitive and Affective Development" (report presented to the Office of Economic Opportunity, 1969, contract B89–4536).
12. Marshall S. Smith and Joan S. Bissell, "Report Analysis: The Impact of Head Start," *Harvard Educational Review* 40 (1970): 51–104; for a reply, see Victor G. Cicirelli, John W. Evans, and Jeffrey S. Schiller, "The Impact of Head Start: A Reply to the Report Analysis," *Harvard Educational Review* 40 (1970): 105–129.
13. Maris A. Vinovskis, "School Readiness and Early Childhood Education," in *Learning from the Past: What History Teaches Us about School Reform*, ed. Diane Ravitch and Maris A. Vinovskis (Baltimore: Johns Hopkins University Press, 1995), 243–264; Zigler and Muenchow, *Head Start*.
14. Zigler and Muenchow, *Head Start*, 10. Max Wolff and Annie Stein, "Six Months Later: A Comparison of Children Who Had Head Start, Summer 1965, with Their Classmates" (New York: Ferkauf Graduate School of Education, Yeshiva University, August 1966)(ERIC no. ED 015025).
15. Maris A. Vinovskis, *History and Educational Policymaking* (New Haven, Conn.: Yale University Press, 1999), 89–114.
16. Richard Elmore, "Follow Through: Decision-Making in a Large-Scale Social Experiment" (Ph.D. diss., Harvard University, 1976); Alice M. Rivlin and P. Michael Timpane, eds., *Planned Variation in Education: Should We Give Up or Try Harder?* (Washington, D.C.: Brookings Institution, 1975); Vinovskis, *History and Educational Policymaking*, 89–114.
17. Vinovskis, *History and Educational Policymaking*, 89–114.
18. Lawrence J. Schweinhart and David P. Weikart, *Young Children Grow Up: The Effects of the Perry Preschool Programs on Youth through Age 15* (Ypsilanti, Mich.: High/Scope Press, 1980); John R. Berrueta-Clement, Lawrence J. Schweinhart, W. Steven Barnett, Ann S. Epstein, and David P. Weikart, *Changed Lives: The Effects of the Perry Preschool Program on Youth through Age 19* (Ypsilanti, Mich.: High/Scope Press, 1984); Lawrence J. Schweinhart, Helen V. Barnes, and David P. Weikart, *Significant Benefits: The High/Scope Perry Preschool Study through Age 27* (Ypsilanti, Mich.: High/Scope Press, 1993).
19. Edward F. Zigler, "Formal Schooling for Four-Year-Olds? No," in *Early Schooling: The National Debate*, ed. Sharon L. Kagan and Edward F. Zigler (New Haven, Conn.: Yale University Press, 1987), 27–44.
20. Information about the gender differences first emerged in Helen V. Barnes, "Predicting Long-Term Outcomes from Early Elementary Classroom Measures in a Sample of High-Risk Black Children" (Ph.D. diss., University of Michigan, 1991). More differences by program participation were discovered at age twenty-seven in areas such as home ownership and earnings of employed males. Schweinhart, Barnes, and Weikart, *Significant Benefits*.
21. Vinovskis, *History and Educational Policymaking*, 74–76.
22. Vinovskis, "School Readiness."
23. Ron Haskins, "Beyond Metaphor: The Efficacy of Early Childhood Education," *American Psychologist* 44 (1989): 274–282.
24. W. Stephen Barnett, "Long-Term Effects of Early Childhood Programs on Cognitive and School Outcomes," *Future of Children* 5 (1995): 25–30; W. Stephen Barnet, "Long-Term Effects on Cognitive Development and School Success," in *Early Care and Education for Children in Poverty: Promises, Programs, and Long-Terms Results* (Albany: State University of New York Press, 1998).

25. Deanna S. Gomby, Mary B. Larner, Carol S. Stevenson, Eugene M. Lewit, and Richard E. Behrman, "Long-Term Outcomes of Early Childhood Programs: Analysis and Recommendations," *Future of Children* 5 (1995): 6–24.
26. "Bush Campaign Proposals on Education," *Education Week* (January 10, 2001), retrieved from http://edweek.org.
27. The National Head Start Association (NHSA), for example, opposes moving the Head Start program from the Department of Health and Human Services. National Head Start Association, "National Head Start Association 2002 Policy Agenda," retrieved from http://www.nhsa.org.
28. Linda Jacobson, "Administration Launches Effort to Boost Early-Childhood Skills," *Education Week* (August 8, 2001), retrieved from http://edweek.org.
29. "An ESEA Primer," *Education Week* (January 9, 2002), retrieved from http://edweek.org.
30. Erik W. Robelen, "Bush Outlines Plan to Boost Pre-K Efforts," *Education Week* (April 10, 2002), retrieved from http://edweek.org.
31. Linda Jacobson and Darcia Harris Bowman, "Early-Childhood Education Advocates Say President's Budget Fails to Meet His Rhetoric," *Education Week* (February 13, 2002), retrieved from http://edweek.org.
32. The Bush administration and the Congress is relying heavily on the reading research and leadership of G. Reid Lyon of the National Institute of Child Health and Human Development (NICHD) at the National Institutes of Health (NIH) as well as on the widely publicized National Academy of Science summary volume on reading research, Catherine E. Snow, M. Susan Burns, and Peg Griffin, *Preventing Reading Difficulties in Young Children* (Washington, D.C.: National Academy Press, 1998).
33. Richard L. Allington, "Troubling Times: A Short Historical Perspective," in *Big Brother and the National Reading Curriculum: How Ideology Trumped Evidence* (Portsmouth, N.H.: Heinemann, 2002), 4. For a scholarly debate over the teaching of reading and math today, see Tom Loveless, ed., *The Great Curriculum Debate: How Should We Teach Reading and Math?* (Washington, D.C.: Brookings Institution Press, 2001).
34. U.S. Senate Budget Committee Task Force on Education, "Prospects for Reform: The State of American Education and the Federal Role," Interim Report (Washington, D.C., 1998).
35. Linda Jacobson, "Ed. Dept., Advocates Clash at NAEYC Meeting," *Education Week* (November 14, 2001), retrieved from http://edweek.org.
36. Linda Johnson, "Administration Launches Effort to Boost Early-Childhood Skills," *Education Week* (August 8, 2001), retrieved from http://edweek.org.
37. On the research needs for Head Start, see Deborah A. Phillips and Natasha J. Cabrera, eds., *Beyond the Blueprint: Directions for Research on Head Start's Families* (Washington, D.C.: National Academy Press, 1996). On the difficulties of the federal government doing rigorous program-related research, see Maris A. Vinovskis, "Missing in Practice? Development and Evaluation at the U.S. Department of Education," in *Evidence Matters: Randomized Trials in Education Research*, ed. Frederick Mosteller and Robert Boruch (Washington, D.C.: Brookings Institution Press, 2002), 120–149; Maris A. Vinovskis, *Revitalizing Federal Education Research and Development: Improving the R&D Centers, Regional Educational Laboratories, and the "New" OERI* (Ann Arbor: University of Michigan Press, 2001).
38. For more on the goals and development of American kindergartens, see Kristen Dombkowski, "Will the Real Kindergarten Please Stand Up?: Defining and Redefining the Twentieth-Century US Kindergarten," *History of Education* 30, no. 6 (2001): 527–545.

39. S. Dianne Lawler, "Kindergarten: New Directions with an Old Philosophy," *Reading Improvement* 25 (fall 1988): 233–236; public school kindergarten provision and enrollment data from National Kindergarten Association, "Important Projects to Help Give Nearly 2 Million Children the Advantages of Kindergarten" (1964), Archives of the National Kindergarten Association, Department of Special Collections, Milbank Memorial Library, Teachers College, Columbia University; National Education Association, Research Division, Kindergarten Practices, 1961 Research Monograph 1962–M2 (Washington, D.C., 1962).
40. Dombkowski, "Real Kindergarten."
41. See Luella A. Palmer, *Adjustment Between Kindergarten and First Grade, Including a Study of Double Sessions in the Kindergarten*, U.S. Department of the Interior, Bureau of Education Bulletin No. 24 (Washington, D.C., 1915).
42. Frances O'Connell Rust, *Changing Teaching, Changing Schools: Bringing Early Childhood Practice into Public Education* (New York: Teachers College Press, 1993).
43. Mary Renck Jalongo, "What Is Happening to Kindergarten?" *Childhood Education* 62 (January/February 1986): 154–160.
44. Ibid.
45. Consortium for Longitudinal Studies, *As the Twig Is Bent: Lasting Effects of Preschool Programs* (Hillsdale, N.J.: Lawrence Erlbaum, 1983).
46. D. R. Entwistle, K. L. Alexander, D. Cadigan, and A. M. Pallas, "Kindergarten Experience: Cognitive Effects or Socialization?" *American Educational Research Journal* 24 (1987): 337–364.
47. This was true into the twenty-first century as well. For a state-by-state breakdown of differences in kindergarten-related policies and provision, see Sara Vecchiotti, *Kindergarten: The Overlooked School Year* (New York: Foundation for Child Development, 2001), 38.
48. Rust, *Changing Teaching*, 8.
49. For a detailed examination of the transition, see Kristen Dombkowski, "Early Childhood in Transition: The Home-to-School Transition in Historical Context" (paper presented at the annual meeting of the History of Education Society, Chicago, October 1998).
50. A 1988 Carnegie Foundation study showed the persistence of these views: 71 percent of parents surveyed wanted more reading and math in kindergarten. Vito Perrone, "The 'Discovery' of the Early Years," in *Shaping the Future for Early Childhood Programs*, ed. Lawrence J. Schweinhart and Leslie de Pietro, High/Scope Early Childhood Policy Papers No. 7 (1988).
51. Maurice R. Berube, *American School Reform: Progressive, Equity, and Excellence Movements, 1883–1993* (Westport, Conn.: Praeger, 1994), ch. 6.
52. National Commission on Excellence in Education, *A Nation at Risk: The Imperative for Educational Reform* (Washington, D.C.: Government Printing Office, April 1983). Though this report focused especially on secondary education, its recommendations for the raising of standards and expectations were deemed applicable to schools at all levels.
53. U.S. Department of Education, *Goals 2000: Educate America Act October 1996 Update*, retrieved from http://www.ed.gov/G2K/g2k-fact.html.
54. National Education Goals Panel, *The National Goals Report: Building a Nation of Learners*, 1997 (Washington, D.C.: National Education Goals Panel, 1997).
55. David Elkind was among the most outspoken and influential of these. See David Elkind, *Miseducation: Preschoolers at Risk* (New York: Alfred Knopf, 1987).
56. Samuel J. Meisels, "High-stakes Testing in Kindergarten," *Educational Leadership* 46 (April 1989): 16–22; Deborah L. Gold, "Mississippi to End Standardized Tests

for Kindergartners," *Education Week* (August 3, 1988), retrieved from http://edweek.org.

57. Note the difference between this view and that of early childhood specialists in the 1930s who believed that ages two through eight were a single category. Sue Bredekamp, ed., *Developmentally Appropriate Practice in Early Childhood Programs Serving Children from Birth through Age 8*, expanded ed. (Washington, D.C.: National Association for the Education of Young Children, 1987), iv.
58. Ibid., 6.
59. Lawrence J. Schweinhart and David P. Weikart, "Why Curriculum Matters in Early Childhood Education," *Educational Leadership* 55, no. 6 (March 1998): 57–60.
60. Snow, Burns, and Griffin, *Preventing Reading Difficulties*, 179.
61. John M. Love, Mary Ellin Logue, James V. Trudeau, and Katherine Thayer, *Transitions to Kindergarten in American Schools: Final Report of the National Transition Study* (Portsmouth, N.H.: RMC Research Corporation, 1992), 5.
62. Ibid., ix.
63. Debra Viadero, "Longitudinal Study Is First to Track Kindergartners," *Education Week* (April 27, 1994), retrieved from http://edweek.org/. For more on ECLS-K, see http://nces.ed.gov/ecls/kindergarten/studybrief.asp.
64. Viadero, "Longitudinal Study."
65. Ibid.
66. Jerry West, Kristin Denton, and Elvie Germino-Hausken, *America's Kindergartners* (Washington, D.C.: National Center for Education Statistics, 2000); Jerry West, Kristin Denton, and L. Reaney, *The Kindergarten Year* (Washington, D.C.: National Center for Education Statistics, 2001); Jerry West and Kristin Denton, *Children's Reading and Mathematics Achievement in Kindergarten and First Grade* (Washington, D.C.: National Center for Education Statistics, 2002).
67. West, Denton, and Germino-Hausken, *America's Kindergartners*.
68. Dombkowski, "Early Childhood in Transition."
69. For general histories of day care in the United States, see Bernard Greenblatt, *Responsibility for Child Care* (San Francisco: Jossey-Bass, 1977); Virginia Kerr, "One Step Forward—Two Steps Back: Child Care's Long American History," in *Child Care, Who Cares?* ed. Pamela Roby (New York: Basic Books, 1973); Lamb et al., *Child Care in Context*; Margaret Steinfels, *Who's Minding the Children?: The History and Politics of Day Care in America* (New York: Simon and Schuster, 1973); Edward Zigler and Edmund Gordon, eds., *Day Care: Scientific and Social Policy Issues* (Boston: Auburn House, 1982); Geraldine Youcha, *Minding the Children: Child Care in America from Colonial Times to the Present* (New York: Scribner, 1995).
70. On World War II child care, see Susan Elizabeth Riley, "Caring for Rosie's Children: Child Care, American Women and the Federal Government in the World War II Era" (Ph.D. diss., University of California, Berkeley, 1996); Anna Smith, "Seeds of Conflict: A Second Look at World War II Day Care Policy" (paper presented at Policy History Conference, St. Louis, May 27–30, 1999); William M. Tuttle Jr., *Daddy's Gone to War* (New York: Oxford University Press, 1993); William M. Tuttle Jr., "Rosie the Riveter and Her Latchkey Children: What Americans Can Learn About Child Care from the Second World War," *Child Welfare* (January–February 1995): 92–114.
71. On domesticity and domestic policy-making, see Karen Anderson, *Wartime Women* (Westport, Conn.: Greenwood Press, 1981); Elaine Tyler May, *Homeward Bound: American Families in the Cold War Era* (New York: Basic Books, 1988); Talcott Parsons and Robert Bales, *Family, Socialization and Interaction Process* (Glencoe, Ill.: Free Press, 1955).

72. John Bowlby, *Child Care and the Growth of Love* (London: Pelican Books, 1953); John Bowlby, *Maternal Care and Mental Health* (Geneva: World Health Organization, 1952).
73. Conference Proceedings Golden Anniversary White House Conference on Children and Youth, March 27–April 2, 1960, Washington D.C.; *Deprivation of Maternal Care: A Reassessment of its Effects* (Geneva: World Health Organization, 1962); F. I. Nye and L. W. Hoffman, *The Employed Mother in America* (Chicago: Rand McNally, 1963); Gilbert Steiner, *The Children's Cause* (Washington: Brookings Institution, 1976).
74. Steiner, *The Children's Cause*, 60–89.
75. "The President's Message to the Senate Returning S. 2007 without His Approval, December 9, 1971," in *Children and Youth in America: A Documentary History*, vol. 3, ed. Robert Bremner (Cambridge: Harvard University Press, 1974), 717–718; Steinfels, *Who's Minding the Children?* 18–19, 187–194; Steiner, *The Children's Cause*, 90–117.
76. For a chronology of this conflicting research, see Michael Lamb, Kathleen Sternberg, and Robert Kettlinus, "Child Care in the United States: The Modern Era," in Lamb et al., *Child Care in Context*.
77. Ibid., 208–215.
78. Personal Responsibility and Work Opportunity Act of 1996 (P.L.104–193).
79. For a thorough discussion of federal child care policy and advocacy from 1971 to 2001, see Sally S. Cohen, *Championing Child Care* (New York: Columbia University Press, 2001).
80. Ibid., 201–214.
81. Information retrieved from http://www.acf.dhhs.gov/programs/ccb.

Chapter 8

The Death of the City

Cultural Individualism, Hyperdiversity, and the Devolution of National Urban Policy

ZANE L. MILLER

SOCIAL SCIENTISTS, even those who oppose the establishment of a national urban policy, tend to equate such a policy with federal government activities. Since about 1950, however, most local, state, and private urbanistic policies have stemmed from the grass roots, appeared in diverse places, and antedated federal government action. The ubiquity of these policies made them national in scope, and they proliferated in starkly varied settings because they derived not from regional, social, economic, or other deterministic forces but from free-floating and popular new ideas about the origin and nature of culture.

Most important, these new ideas fostered the notion that individual Americans should define their own ways of living and dying, the logic of which led to demands for a hyperdiversity of choices in the selection or design of personally satisfying social and physical environments. Since the mid-twentieth century, in short, this culturally individualistic mental apparatus has transformed American views about who should arrange our civilization and how. Thomas Lynch, for example, an undertaker, has pointed out the effects of this culturally individualistic mental apparatus on the conduct of his business. The apotheosis since 1950 of choice and diversity, he argued, eroded the influence of expertise and traditions. As he put it, his predecessors "occupied a world of black and white, custom and tradition, moral certainty," values that led them to offer a narrow range of casket types and to tout all of them for their virtues of "protection and permanence." But now, Lynch and his colleagues "see our rights and wrongs as relative; we roll our own orthodoxies," including the view that customers should choose their own burial styles. As a consequence, writes Lynch, "choice" and "options" rank as the buzzwords of the business, everything is "customized," and "the trade is brisk in wakes and funerals that offer a personalized touch."

> My father's generation did copper and concrete and granite memorials. We do biodegradables, economy models and ecofriendly cyberobsequies. He sold velvet and satin crepe interiors. We do urns that look like golf bags and go to cemeteries with names like golf courses. You can buy a casket off the Internet, or buy plans for a self-built "coffin table" or one that doubles as a bookshelf until you "need" it. There's a push for "do it yourself" funerals. . . . Cremated remains can be recycled as memorial kitty litter, sprinkled on rosebushes, mixed with our oil paints to add texture to fresh masterpieces.[1]

So for funerary policy, so, too, with national urban policy, where attempts to realize the post–1950 version of diversity made individual choices, options, customized, personalized, and specialized keywords as a new way of thinking about and living in the city took hold.[2] After midcentury the making of urban policy increasingly rested on an understanding of the city stemming from a view of American civilization among social scientists and others that took individuals as the basic units of society and that apotheosized the right and obligation of individuals to define or choose and change lifestyles unfettered by outside restraints. This culturally individualistic stance generated an antiexpertise and antistatist animus on the political right, left, and center that prompted federal, state, and local governments to push down as close to the grass roots as possible the previously "elitist" tasks of defining urban problems and proposing, selecting, and establishing guidelines for implementing solutions to problems.[3] And since liberated individuals at the grass roots differed in their lifestyle preferences, they often stitched together self-constructed advocacy institutions of like-minded persons as agencies through which to establish policies that would help them realize and/or define their lifestyle choices.

At first, citizen empowerment advocates viewed government as an ally. But increasingly in the 1960s and after they regarded government as a problem, an obstacle to the achievement of their aims. Nonetheless, their culturally individualistic ethos made personal identity politics (and changing identities) a hallmark of American urban life and policy throughout the second half of the twentieth century. Some individuals, for example, opted for a lifestyle that centered on making a lot of money by fostering economic development and metropolitan growth, while others preferred a style that de-emphasized materialism and advocated a no- or slow-growth policy to reduce wear and tear on the environment. Others who selected a lifestyle that involved religion in public schools clashed with those who adopted an identity more comfortable with a strict separation of church and state. And those who sought the ways, means, and space necessary for a full and free expression of their self-defined roles as feminists, homosexuals, members of this or that ethnic or racial minority, homeless people, or panhandlers often ran into opposition from people who worried that the legitimation of such behavior would push to the margins of society those self-styled "normal" individuals who defined the choice of "straight" social homogeneity as mainstream and the marginalization if not the elimination of otherness as a necessary adjunct of that pursuit.[4]

The "straight" social homogenizing category, however, comprised a small and diverse cadre of outsiders, including anti-Semitic, antiblack, and antihomosexual backlashers. Their agenda ran counter to cultural individualism, but their ideas excluded from their ranks most self-styled Christian conservatives. The latter expressed their concern about urban life and the direction of American civilization in terms of their sense of the decline of moral values as manifested, for example, in the public flaunting of sex, violence, obscene language, and homosexuality; in the high rates of crime and in drug and alcohol abuse among youths; and in the proliferation of pornography, broken homes, and the resort to abortions (conceived of as murder). Indeed, many conservative and most moderately religious people in big-city neighborhoods, suburbs, and small towns disliked but did not condemn such things on moral grounds (except perhaps abortion) for public policy purposes because they upheld cultural individualism, freedom of speech, and tolerance with respect to lifestyle practices that fell outside their private view of the moral realm.[5]

In this relativistic context the culturally individualist ethos encouraged elected and appointed officials, including judges, to duck issues by deferring to "community standards" and to promote maximum feasible participation in policy-making as ways to defuse lifestyle political conflicts. But these processes often proved tediously contentious and litigious and provided the basis for the proliferation of citizen-produced legislation via referenda and/or the revising of home rule city charters because lifestyle advocacy groups regarded their pursuit of self-actualization as a civil right, an issue of uncompromisable high principle.[6] In addition, some such groups worried that their members might exercise their right to identity choice by switching to another lifestyle, a concern that often provoked attempts to enlist governmental aid in geographically constricting (segregating) or banning alternative lifestyles that a particular group regarded as dangerously seductive in a society that defined the right to make and change lifestyle choices as a key to self-fulfillment, psychological health, and social and civic happiness.

Yet the intense and intensifying pluralism of metropolitan areas generated by cultural individualism stabilized the persistence of hyperdiversity as a durable feature of urban life. It did so in large part by challenging those who believed they possessed an ultimate truth and a mission to bring this gift to others (such as evangelical Protestants, for example) to engage tenaciously on their home grounds in an irresistible though unwinnable lifestyle sweepstakes that pitted them not only against alternative true believers but also against various advocates of toleration on relativistic grounds.[7]

The human construction of conflicting cultural groups, the devolutionary impulse in policy-making, and other familiar trends, tendencies, problems, and issues in urban life after 1950 may be more clearly understood and more appropriately assessed in light of another midcentury phenomenon, the death of the idea of the city as a culture-generating social and physical environment. During the second quarter of the twentieth century, that is, social scientists, city planners, and other government officials contended that groups, not individuals, formed the basic units of American life and that individuals acquired their identities and lifestyles from the fact of their membership and participation in the life of the group. These policy-makers also tended to argue that groups acquired their identities and ways of life from their history and experience in particular places (social and physical environment). According to this view, the culture of a group could stem from its history and social experience in a territorial place, such as Appalachia, the South, the West, a foreign nation, or a particular neighborhood in a particular city. But in this vision the culture of a group could also derive from the place of a group on a social scale, measured sometimes in terms of class, or occupation, or gender.

Urban policy-makers of the 1920s and 1930s also tended to see the culture of groups as deterministically dynamic, as gradually changing because of the influence of "forces" or "factors" beyond the control of group members. Usually the policy-makers argued that the migration of groups produced intergroup contact, competition, and cooperation from which flowed economic and technological changes that slowly altered the culture of groups, most often in a process that moved some groups from a position of isolation in rural areas to a position of intense intergroup contact in the city. In this view, moreover, the experience of urbanized groups within the city also stemmed from the influence of similar dynamic factors, in this case intramural migration and interaction leading to economic and technological change. In this scheme of things, groups with the most experience in city living flourished because of the evolution of their cultures in step with the phases of economic and technological evolution, and such groups retained their distinctive cultures while modernizing as the city modernized. Those groups with less experience in the city "lagged" behind economically and technologically, and the most recently arrived groups lagged so far behind that they stood in danger of entrapment in slums that stripped them of their culture and their attachment to the social and civic life of the city.

This understanding of urban life meant that the United States, its regions, and its cities suffered and always would suffer from problems. Yet the urban policy-makers of the 1920s and 1930s faced the future cheerfully, for they defined cities as benignly dynamic because they concentrated the greatest variety of groups under conditions of close proximity and intense interaction. These factors seemed like good things because the urbanists believed that such conditions drove a form of economic and technological change that in the long run yielded, or could be managed so as to yield, improved standards of living for all groups and ever more inclusive democratic practices.

The deterministic cultural pluralism of the 1920s and 1930s, in short, identified "progress" with the process of urbanization and regarded it as inevitable so long as the great variety of groups interacted in a competitive and cooperative manner that kept the social and civic peace. And this seemed likely, for the urbanists concluded that sophisticated managers of intergroup relations could persuade individuals to pursue self-fulfillment by behaving in socially and civically responsible ways, especially by placing the welfare of the whole ahead of group interest. The key to such good citizenship rested in the cultivation of intergroup understanding and tolerance while at the same time fostering the residential segregation of groups as a means of keeping the peace among groups that might otherwise clash in social and economically disruptive ways, leading to uncompromising political discord and even violence. Some urban policy analysts, however, regarded such segregation as temporary, as a step in the process by which astute urban design and management would render all groups economically and technologically competent, a situation in which they might be guided to share traits while retaining their sense of their group identity. Such a cosmopolitanizing process, its advocates contended, would make easier the keeping of the social and civic peace, facilitate the advance of democracy, promote the public welfare, and gradually render old fashioned and obsolete the practice of residential segregation.

After midcentury, however, more and more urban policy analysts rejected cultural engineering. By then it seemed evident to many that the gradualistic policies of the 1920s and 1930s had not produced a new era of intergroup tolerance and understanding, social harmony, economic progress for all groups, and a more inclusive democracy. But the rejection of cultural engineering also derived from a general revolt against the notion of the social (and biological) determination of culture on which cultural engineering rested. This now seemed a dangerous mode of thinking because of its association in Europe with fascism, Nazism, and communism and in Asia with the Shintoism of Imperial Japan. These ideologies differed in detail, of course, but they struck many influential Americans in the 1940s and 1950s as totalitarian cultural engineering movements that resembled the urban variety of that practice in America because they, too, projected cultural blueprints of the future and preached sacrifice for the welfare of the whole as necessary for promoting and defending the public interest. In addition, such ideas in the context of the Cold War seemed deplorable because they might create a "soft" totalitarianism, a democratic but conformist mass society incapable of appreciating cultural diversity in any form.[8]

As a matter of practice, the revulsion against deterministic views of the origins, nature, and future of group cultures undermined efforts to create a national urban policy by the federal government and to continue metropolitan master planning as defined in the 1920s.[9] Both endeavors involved the use of "grand theory," the assessment of past and present conditions of "given" group cultures in the belief that such studies could predict the direction of history and provide the basis for the development of policies compatible with that direction. This belief came under fire in the mid–1930s on relativistic grounds from Charles A. Beard, who remained nonetheless quite willing, despite his embrace of the relativistic principles of uncertainty and/or unknowability about the long run, to make a judgment about the future as "an act of faith." Beard chose social democracy, but by midcentury "act of faith" policy-making seemed a dicey proposition on which appeals for support for undemocratic causes might be couched in zealously ideological and fanatically "romantic" terms. The more profound relativism of the age of cultural individualism suggested the utility of sacrificing both rational and irrational "grand theory" for "middle range theory," which suspended both rational and irrational judgment about the direction of it all and provided for a pragmatic policy of dealing with particular problems on the basis of careful assessments of the opinions of self-constructed groups of individuals.[10]

The revolt against determinism, in short, suggested the utility of empowering people to define their own lifestyles and cultures, an apotheosis of the autonomous/liberated individual and of seeking self-fulfillment through self-actualization that cast a dim light on crucial ideas of the past in urban politics.[11] The promotion of self-development for psychological rather than the political good rendered suspect both self-styled civic "experts" and governments as potential repressors of the right of self-determination in the name of the welfare of the whole. The same impulse regarded intergroup confrontation rather than compromise as the honorable method by which individuals might establish for themselves a satisfying way of life. And cultural individualism's emphasis on securing psychological serenity through self-actualization in a universe of self-defining individuals made diversity and choice prime keywords in the new era.

Cultural individualism's psychologizing of American society and civic life also transformed city-building aspects of urban policy.[12] It sparked efforts to create and re-create a dazzling variety of social and physical environments in each region of the country, in central business districts, in inner- and outer-city turfs, and in suburban tracts of metropolitan areas. All these places no longer seemed generators of a "given" and distinctive culture. Instead, they now looked like placeless blank slates capable of development socially, economically, and culturally in any direction that old or new residents might wish to take them. All of them as a consequence became a potential arena of conflict among self-constructed advocacy groups, all of which agreed only on the right to contest which "community values" might become ascendant in the locality in question.[13]

The combatants proved especially adept at persuading governments and businesses to use such technological innovations as jet planes, expressways, photoduplication equipment, electronic communications systems, and air conditioners in ways that revolutionized long-standing regional patterns, especially the nature of cities in the South and Southwest. Here the new mode of thought enabled aggressive and imaginative boosters to shake off their sense of their regions as economic colonies of the North with a "given" culture that drastically narrowed the range of imaginable visions of the future. Now cultural individualism created a spirit of urban imperialism that made both regions national leaders in rates of metropolitan growth. The expansionists' tactics included the adding of new attractions to old ones, such as climate, scenery, and the historic particularities of various cities, as a way of retaining population and attracting "outside" people. Imperialistic zeal spawned new national and international corporations that planted their corporations in Sun Belt downtowns and suburbs and set off campaigns to buy out nationally famous northern businesses, such as New York's ritzy Saks department store and the Milwaukee-based Carson Pirie Scott.[14] And in this context Sun Belt city boosters summoned the imagination and resources to develop and sustain not only minor but also major league arts institutions and events, such as the Spoleto Festival in Charleston, South Carolina, and either to establish or purchase and bring home big league professional sports franchises in basketball, baseball, and football. In the process, Sun Belt city boosters broke the monopoly in these fields long held by the urban-industrial heartland, a region that contained the most dynamic cities of the late nineteenth and twentieth centuries in a stretch of territory reaching from Boston and Baltimore in the east across the Great Lakes corridor and upper Ohio Valley to Cincinnati, St. Louis, Chicago, and Minneapolis.

Some urban heartland politicians so resented the success of the upstart South and Southwest in attracting people, jobs, and big league leisure institutions from heartland metropolitan areas that they threatened to cut off or reduce federal subsidies to those areas, which received more in federal expenditures than they sent to Washington in tax dollars. The election in 1976 of a Georgian, Jimmy Carter, as president of the United States intensified both urban heartland angst about the interregional competition and the campaign by some heartlanders to level the playing field. The head of the Dallas Chamber of Commerce tagged the contest a second "civil war," but this time, he asserted, "we're going to win."[15] The competition indeed persisted, and the prediction of an ultimate Sun Belt victory seemed an even better bet in the 1990s when another southerner, Bill Clinton, moved into the White House. The odds improved even more when southern and southwestern Republicans captured all the major leadership positions in the U.S. Congress, which fell into the hands of the GOP in large part because voters in the states of the former Confederacy abandoned in the decades after 1960 their once "solid" Democratic allegiance and adopted new political identities as Republicans as the best way of realizing their new self-selected lifestyles. Some former Sun Belt Democrats, however, identified themselves as independents and wavered in particular elections, sometimes working and/or voting for GOP candidates and sometimes for political mavericks, such as Texas billionaire Ross Perot, in the late 1980s and early 1990s.

From this political perspective, the first major step in the switch to the idea of cultural individualism in urban policy may be dated to the Democratic national convention of 1948, which for the first time put a strong civil rights plank in the party platform. At that meeting Mayor Hubert Humphrey of Minneapolis stunned the delegates with a passionate antigradualism speech. Humphrey spoke, he said, with no single racial or religious group or region in mind, and he took his cue from the convention's keynote speaker, Kentucky senator Alben Barkley, who had reminded the convention that the equality clause of the Declaration of Independence applied to black, red, and yellow people; to Christians and Jews; to Protestants and Catholics. But Humphrey pitched his proposal as a moral issue that required immediate attention because the United States could not afford any longer to compromise as it faced the challenge to its international leadership by atheistic Communism, "the world of slavery." The time had come, Humphrey went on, to move "towards the realization of a full program of civil rights for all . . . human beings . . . , people of all kinds," so that they could be "truly free and equal" to use their "freedom and equality wisely and well." The high point of his address came when he invited the delegates to put individual rights beyond the reach of governments by stepping "out of the shadow of state's rights to walk forthrightly into the bright sunshine of human rights."[16]

This event ignited a reaction that may be seen as the start of the cultural liberation of the "Solid South" from its identity as deterministically Democratic because of its unique history and sociodemographic composition. The party's new stance on civil rights as an uncompromisable moral issue so angered Governor J. Strom Thurmond of South Carolina that he and thirty-eight other southern delegates walked out of the convention in protest. In an act of political revenge, Thurmond then sought to pull down the Democratic presidential ticket by running for the White House himself on a "Dixiecrat" slate. Thurmond lost, but he still simmered about the civil rights question and finally, as a U.S. senator, completed his political makeover by joining the GOP, which welcomed with open arms both Thurmond and other white southern Democrats who decided to join the GOP, a sign of the undermining of party loyalty and discipline in the new age of the independent candidate, independent officeholder, and independent voter.

But others responded differently. Arthur M. Schlesinger Jr., an organizer of the liberal anti-Communist organization Americans for Democratic Action, tried to commit it and the nation to the new approach to civil rights, which he defined in 1949 as a commitment to "the ultimate integrity of the individual," a view that underlay both the gradual easing between 1943 and 1965 of restrictions on immigration and the culturally individualistic transformation of city planning policy.[17] That policy may be dated to the Federal Housing Act of 1954, which mandated, ambiguously, citizen participation in planning for cities that chose to apply for funds for urban renewal projects.[18] Some cities responded to this mandate merely by setting up citywide citizen committees to react to and comment on proposals put forth by urban renewal officials. But others, such as Baltimore and Cincinnati, read the citizen participation mandate as an opportunity to move planning in a new and nonauthoritarian direction by recognizing the autonomy of neighborhoods and by recruiting residents to participate in the definition of the social and physical problems to be addressed, in devising solutions for those problems, and in implementing plans to create a personalized and customized neighborhood as a remedy for the problems. This proved a cumbersome process, and the first such project in Cincinnati took much longer than expected because of controversies that pitted citizens against one another and the bureaucrats. It also proved unsatisfactory, according to the federal Housing and Urban Development officer who evaluated the project. She complained, however, not about the inefficiency of the process or the substance of the plan but about the failure of city officials to engage residents early enough, seriously enough, and long enough to guarantee genuine neighborhood autonomy.[19]

But the great avalanche of culturally individualistic federal, state, and local cultural empowerment laws and programs came during the 1960s and 1970s in forms designed to provide equal access for all citizens to choices among rights, responsibilities, opportunities, privileges, experiences, and neighborhoods. These include measures that offered individuals choices in jobs, education, housing, and other areas of discrimination in the public and private sectors that affected not only all regions, races, religions, and ethnic groups but also women, gays, senior citizens, single parents with children, and people with disabilities. The federal government in addition not only facilitated but sometimes mandated the maximum feasible participation of citizens in defining urban problems, developing remedial programs, disbursing funds, implementing projects, and evaluating the policy-making processes and their consequences in a variety of federally financed projects under the Equal Opportunity Act, the Model Cities Act, general and special revenue sharing, and community development block grant programs.[20]

Local officials used these federal subsidies in a variety of particular projects, but they invariably aimed at the diversification of social and physical environments as a means of expanding choices, options, and experiences for residents. These programs included efforts to make central business districts as lively by night as by day through the creation within them of heterogeneous residential land-use components and entertainment facilities. Much of this money also went into attempts to expand neighborhood options by creating and sustaining socioeconomically and racially mixed neighborhoods as a supplement to the larger number of racially segregated inner-city, outer-city, and suburban living venues.[21] Federal and other governments also underwrote the diversification of the suburban landscape by boosting projects that mixed white Catholics and Protestants of various ethnic backgrounds and that yielded "edge cities," outlying places for industry, offices, commercial establishments, and entertainment facilities that offered an alternative residential setting for people who disliked living in a big city, or in purely residential suburban subdivisions, or in subdivisions containing just a shopping mall but no significant amount of other land uses.[22]

This same concern for preserving and/or expanding the variety of available neighborhoods sparked worries that urban redevelopment activities in the

1950s and 1960s, including expressway construction and the razing and modernization of physically "blighted" old neighborhoods, would drastically reduce the range of available cityscape options to those residents who preferred to revitalize such localities in ways that preserved their neighborhoods' venerated physical attributes. These worries inspired intense lobbying that produced in 1966 the federal Historic Preservation Act, which helped shift the focus in historic preservation from authentic restoration and national patriotism to the "adaptive reuse" of rehabilitated old structures and the recognition of locally important individuals and events for the purpose of making the commonplace physical legacy of the past "a living part of our community life and development in order to give a sense of orientation [as in identity] to the American people." The act created a register of such places and in culturally individualistic fashion authorized not experts in history or culture but any individual or individual organization to nominate buildings or districts for listing on the register and (as amended) permitted individual property owners unhappy with that step to veto it in advance of the listing. And while the act did not prohibit the demolition of historic resources, it sought to deter their rash destruction in projects involving federal funds by instituting a review process to assess the impact of such projects on listed or potential historic sites and to seek ways of voluntarily mitigating any adverse effect, not just demolition, on such resources.

Ardent historic preservationists acquiesced in this tepid federal effort to protect historic sites and neighborhoods because they possessed a more potent local remedy, the use of the policy power through the development of historic preservation zoning districts as overlays to other land-use regulations in the area. Preservationists argued that visual blight threatened the public health, safety, and welfare and justified the "taking," without compensation, of the unfettered use of private property that qualified as historic under the criteria of the zoning ordinance, usually a copy of the criteria set down in the National Register legislation. Most local historic district legislation set down guidelines, not mandates, for rehabilitating old buildings and cityscapes and allowed property owners who wanted to tear down a designated site to do so after showing that no economically feasible use of it could be found.

Historic preservation zoning ordinances ordinarily also provided amply for citizen participation in the nomination process. Characteristically, citizens helped define district boundaries and design regulations, after a board of volunteers appointed by the mayor or the city manager held hearings on all nominations, sent those it approved to the city planning commission, which held more hearings and forwarded its decision to the city council, which held yet another round of hearings. This cumbersome process made it fairly easy for a well-organized and disciplined antipreservation neighborhood or organization to block the local listing by giving them three shots at killing the proposal before three different sets of officials.

The culturally individualistic urge to customize and personalize social and physical environments produced not only historic district zoning but also a less well known orgy of land-use regulation alterations that fattened zoning codes ponderously. Cincinnati's city council, for example, revised in 1963 the city's first comprehensive zoning code (1924), then made 627 changes in the new version between 1963 and January 1983 that yielded the publication of a 435-page revised zoning code.[23] In the Queen City and elsewhere, land-use regulators laid out environmental districts of various kinds to protect hillsides and hillside views, areas of high public investment, and neighborhood business districts that had adopted design plans. And the creativity did not stop there. Many cities adopted planned unit development (PUD) zones to permit low-rise apartment and condominium construction at relatively high densities when embellished with green-space features, and several either adopted or considered the creation of sex shop and strip joint zones, often after "cultural war" attempts to run such businesses out of town proved unconstitutional, as in the incidents that made Larry Flynt rich and famous. And in Portland, Oregon, the nation's most imaginative land-use regulators even came up with a crime abatement zone that banned from such neighborhoods nonresidents arrested on drug or prostitution charges in such protected spaces.

In the 1980s and 1990s, however, those who opposed the governmental initiatives of the 1960s and 1970s to promote options, choices, diversity, and specialization began to assert that private sector institutions could more effectively promote the achievement of those ends. As one close observer of the city scene in the 1980s and 1990s has noted, the recent drive for customization and personalization in urban design manifested itself quite spectacularly in privately developed supermalls in central business districts, airports, and the outer suburbs. Builders of these places decided to equip them not with the wall-less departments that promoted the idea of a seamless interconnection, coherence, and unity of the whole in the way common to early-twentieth-century department stores. Instead, they created within the malls enclosed shops, restaurants, and salons with decorated facades and separate and lockable doors to emphasize each business's specialness and the autonomy of its culture as devised by their managers, whose employees often relinquished temporarily their preferred identities to don costumes reflective or indicative of the business's style and its particular theme.[24]

Supermalls in competing sites and cities, moreover, struggled feverishly in the specialization sweepstakes to accommodate or stimulate the peculiar tastes and modes of self-expression among customers engaged in identity construction, deconstruction, and/or reconstruction. The supermalls became prime venues for customized shopping because they provided sufficient foot traffic to support stores working from very narrowly defined "product bases, such as teddy bears, pretzels, or items with a rainbow motif." And supermall magazine racks also aimed to "feed a personal monomania" by selling single-theme subscription magazines catering to the most bizarre tastes, ranging from "the satire of violence (*Murder Can Be Fun . . .*)," to "pet worship (*Animal Review*, the fanzine of herbivorous youth)."[25] A student of the phenomenon noted that the purchasers of sneakers in one supermall now could choose between a broad variety of specialized shoes—running, tennis, squash, basketball, walking, aerobic, and cross-training shoes—and that they shopped for these goods at gender- and age-specialized stores, such as Foot Locker for males, Lady Foot Locker for females, and KidsFeet for children.[26]

In such big-city, suburban, and airport supermalls people with varied lifestyles mixed together as individuals headed to the specialty shops where they joined others who shared a particular taste to form a temporary community of consumers. But catering to the culturally individualistic drive for specialization and personalization led to other forms of segmentation that reduced drastically the public interaction of diverse people regarded in the 1920s and 1930s as a major and desirable characteristic of urbanism. The popularity of nonbizarre special-interest magazines picked up in the 1950s and 1960s and mounted through the rest of the century, a movement that drew isolated readers from general-interest publications where they might visit other worlds and other lifestyles. The same thing happened in radio and television as the proliferation of broadcasters sapped general interest programming by targeting narrower and narrower "niche" audiences with specialized tastes. The advent of videotape movies accelerated the isolating trend, while personal computers made it possible not only for members of self-constructed groups to talk exclusively with one another but also facilitated the creation of home offices that eliminated the need for workers to leave their houses and travel through the city among different people to a job in a different social and physical environment.[27]

People in and after the 1970s who objected to government action for the empowerment of individual city dwellers in their pursuit of cultural individualism during the 1950s and 1960s now applauded the supermalls and other examples of culturally individualistic private sector consumerism in urban areas during the 1980s and 1990s.[28] But they did not stop there.[29] They now denounced many of the urban initiatives for cultural individualism of the 1950s and 1960s as failures and sought to dismantle them in the name of other and allegedly more effective ways of liberating individuals and individual corporations for the more efficient pursuit of cultural individualism. In the process, federal, state, and local governments adopted positions that permitted, encouraged, or facilitated (and sometimes mandated) private sector actors to specialize, customize, personalize, or diversify in their efforts to accommodate and/or stimulate culturally individualistic behavior. The domestic policies of President Ronald Reagan, a former actor and therefore an adept at adopting and switching identities, epitomized the slogan that dominated urban policy in the 1980s and 1990s: government isn't the solution, it's the problem.[30]

But the antigovernment tendency in national urban policy may be traced to the late 1960s and early 1970s when Congress balked at attempts to move significantly beyond assuring the civil rights of African Americans by establishing a program to shore up black families and by using income tax revenues to pay welfare clients a no-strings-attached amount of money to free them from their dependence on social workers and to help them, especially African American males, to marry and sustain a family. Part of the opposition to these and other new governmental attempts to culturally empower African Americans rested on the new view of ethnicity as a self-constructed identity that enabled white immigrants and their children in the nineteenth and twentieth centuries to work their way out of their old inner-city neighborhoods without the aid of tax-supported and government-specified ethno-centered programs. To advocates of this view, African Americans seemed the most recent of urban newcomers who would benefit from a policy of "benign neglect" that cut back on special government programs to foster African American empowerment while also requiring state and local officials to scatter across the face of the metropolis subsidized housing for the poor rather than concentrating it in certain neighborhoods.[31] According to the immigrant analogy, such benign neglect would liberate African Americans to lift themselves by their own bootstraps and, if they chose, to adopt a lifestyle congruous with living in one of the variety of neighborhoods outside the ghetto.[32]

President Jimmy Carter avoided the rhetoric of benign neglect and embraced the establishment of a new kind of public-private partnership as a strategy for dealing with urban poverty problems in the age of cultural individualism, hyperdiversity, and the devolution of public authority. The public-private partnership idea gained momentum during the 1950s and 1960s with government funding for not-for-profit neighborhood housing and community development corporations, a practice that persisted into and beyond the 1970s. But in the 1970s Carter added a new dimension to the practice by encouraging the voluntary establishment of "self-help" partnerships between local governments and business corporations for the purpose of assisting cities containing a disproportionate share of poor residents. The new initiative came in the form of the Urban Development Action Grant Act (UDAG) and the Housing Development Action Grant Act (HODAG). Under UDAG, interested local governments and businesses worked out building programs to generate construction and service jobs, especially for minorities, and applied to the federal government for money to subsidize a variety of projects, almost all of which consisted of central business district diversification schemes, including particularly waterfront redevelopment projects featuring aquaria, historic preservation districts, theme parks, hotels, shops, offices, and sometimes market-rate housing. Under HODAG, interested local governments and businesses worked out schemes and applied to the federal government for money to facilitate the creation of commercial housing that incorporated space for a diversity of socioeconomic groups, including the reservation of a small percentage of the units for minority persons (a small percentage so as to reassure flighty whites about the stability of the venture).

Some of Carter's advisers also took a self-help approach to the post–World War II flight of jobs from eastern and midwestern industrial cities to the South, Southwest, and Far West, a problem discovered by social scientists in the 1970s and popularly described as a shift of economic vitality and political clout from the Rust Belt to the Sun Belt. A Carter administration national urban policy report argued that market forces stemming from individual lifestyle choice fostered this shift. In this calculus, growth and decline ranked as a neo-Darwinian (culturally individualistic instead of deterministic) aspect of the integral process of urban life. And from this calculus the urban policy conclusion seemed obvious: unskilled Rust Belt workers should move to the Sun Belt and adopt a Sun Belt lifestyle, a process the federal government might facilitate by subsidizing their relocation costs. Carter and Congress, however, adopted a policy of benign neglect on this proposal, a policy choice that underwrote the shift without spending a federal penny by advising workers and others to vote with their feet in favor of the great migration.

The canonization of the private sector as the prime patron of cultural individualism gained ground in the 1970s and led President Ronald Reagan and his followers in his cabinet, in Congress, and at the state, county, and township levels and in city and suburban municipal buildings to promote a variety of ways to unleash a private sector attack on urban social, economic, and physical problems. They eliminated government programs and promoted "supply-side" economic strategy tax cuts in the hope of stimulating economic growth, raising government revenues, and broadening job and business opportunities to enable more individuals in metropolitan areas to gather the resources necessary to define or redefine and practice their preferred lifestyles. They deregulated industries and commercial activities on the grounds that more individual city residents should have the chance to exercise their right of participation in production and selling and on the grounds that competition among such urban entrepreneurs would yield more firms to offer more and more diverse products and services to more and more specialized markets at lower prices.[33]

The deregulation mania of the 1980s and 1990s applied in addition to a variety of big business corporations, many of which used their liberation to engage in corporate mergers on an unprecedented and sometimes international scale. Many of them also used their liberation as a pretext for relaxing their sense of locational obligations and/or loyalty to particular cities, states, or even the United States so they could more effectively pursue opportunities to promote globalistic and culturally individualistic consumerism in remote parts of the world on the assumption that even "alien" people, if given the chance, would embrace the pursuit of self-fulfillment through self-definition, self-actualization, and self-expression, including the expression of fondness for fast-food chain stores and for sex, violence, and vulgar language in various media entertainment forms while indulging also customs and tastes selected from their immediate milieu.[34]

When cutting taxes and deregulating in the name of cultural individualism proved impossible, Reagan and his followers tried to eliminate or reduce the budgets of offensive agencies and programs. These slashes hit a variety of targets, including not only federal legislation designed to resolve urban social and economic problems but also the National Endowment for the Arts, the National Endowment for the Humanities, National Public Radio, and the Corporation for Public Broadcasting. All of these arts and entertainment institutions took shape in the 1950s and 1960s principally because urban-based lobbyists wanted the federal government to provide additional leisure options to an audience that felt that commercial and private patrons overlooked the tastes of a small and intellectually sophisticated segment of society. But the chief supporters and beneficiaries of these programs, people who lived in affluent outer-city neighborhoods and suburbs, found themselves under heavy fire in the 1980s and 1990s on several grounds. Some critics described these agencies as tools of selfish elitists who wanted to embellish at public expense their personal lifestyles. In addition, these allegedly elitist big-city institutions sometimes underwrote museum exhibits, public art or architecture, literary or theater performances, histories, or documentaries that some regarded as immoral or as legitimizing other activities and views deemed unacceptable on patriotic grounds by some self-constructed groups or because they offended the taste of such groups.

Governments responded after the 1970s to cultural individualism's choices, options, and segmentation mantra not only by cutting programs and budgets but also through the "privatization" of many urban services by letting private sector actors carry out tasks once handled by public employees. As a consequence of this movement, a wide range of urban services in many cities around the country passed from the public to the private realm, including garbage collection, prison construction and management, and the operation of municipal golf courses, hospitals, and parking facilities. By the 1990s proponents of privatization had taken aim also at big-city public schools by pushing for the creation of government subsidized voucher programs that would enable parents and their children to select the nonpublic school of their choice. In many cities, public school boards reacted to the voucher challenge by letting parents and their children decide if they wanted to attend so-called charter schools under public auspices but featuring specialized programs different from those in the regular public schools (which themselves had created with the aid of federal funds in the 1950s, 1960s, and 1970s unprecendentedly diverse course offerings that they struggled to sustain in the face of sharp cuts in federal education expenditures during the 1980s and 1990s).[35]

Privatizers also cast skeptical eyes on federally supported public housing projects. The first wave of these projects dated from the 1930s and 1940s and produced in large cities around the country so-called community development housing, large residential sites that included design elements, civic and shopping amenities, and social supervision for the purpose of manipulating the culture of the projects' low-income residents to make them capable of upward mobility, after which the authorities moved them out into established neighborhoods and took on a new batch of critically needy low-income persons. These projects seemed too conformist, almost totalitarian in the 1950s, and reformers sought to improve the community development sites by downplaying their cultural manipulation aspects and by abandoning plans to create additional ones in favor of giving low-income people a greater range of neighborhood choices by scattering small apartment buildings in privately developed residential areas. The scatter-site scheme, of course, encountered stiff resistance during the 1950s from residents of targeted localities who opposed such a change in *their* neighborhood of choice. And that resistance sharpened in the 1960s and after, as expressway construction and slum clearance programs forced public housing authorities to fill both the large-scale projects and the inadequately small number of scatter-site apartments with desperately poor people, including very large numbers of African Americans uprooted from old inner-city areas as the slums fell to road construction and redevelopment.

Appalling conditions in these places, and especially their failure to provide tenants with choices in housing styles and neighborhoods, kept alive the search for alternative yet culturally individualistic solutions to the problem of providing affordable housing for the urban poor in neighborhoods of their choice. The first of these, the so-called Section 8 housing program, provided federal subsidies to take housing authorities out of the business of managing both the community development and scatter-site public housing projects. Under Section 8 of the Federal Housing Act, housing authorities lined up market-rate landlords in a broad variety of neighborhoods across the face of the metropolis and provided them with low-income tenants supplied with federally backed vouchers to make up the difference between what the poor tenant could afford to pay and what the landlord charged for rent.

Section 8 subsidized housing survived into the 1990s, though not on a scale satisfactory to its strongest supporters. By that time, however, public housing authorities decided that they might be able to redeem their old and large community development projects in culturally individualistic ways, especially if they could retain or expand the stock of Section 8 housing in outlying neighborhoods. For this purpose they persuaded the Clinton administration and Congress to put up $2 billion for a program to tear down in twenty-five cities all or parts of the now infamous midcentury projects and to replace them with a mix of subsidized and market-rate dwelling units along with guarantees that any poor resident of remaining project housing who wished to stay on would not be involuntarily moved out of the new environment. And the housing authorities sought not only income and racial/ethnic mixing in these new subdivisions but also the provision of front porches, townhouses, small apartment houses, and free-standing homes set close to one another and to narrow streets lined with broad sidewalks for the purpose of giving residents another choice, the chance to interact in a neighborly fashion with other residents.[36]

The most recent major step in privatization, welfare reform, especially for the inner-city poor, may be the most dramatic example of that phenomenon. President Clinton pledged to "end welfare as we know it," by which he meant welfare as reformulated in the 1950s and 1960s.[37] In those decades state and federal governments sought to help the poor in their quest for self-definition, self-actualization, and self-expression by increasing benefits to leverage them out of deep poverty and by devising ways of letting the poor participate effectively in defining their own problems and devising solutions, especially governmental ones, to those problems. This led to rising expenditures for welfare but no significant reduction in the welfare roles in metropolitan areas.

After the 1970s this form of welfare fell into general disrespect as critics decried the cost in dollars and increasingly pointed to welfare's failure to relieve the urban poor of their dependence on government bureaucrats and private social workers for guidance in lifestyles, guidance that seemed to have failed. Instead of liberating the poor, participatory welfare as shaped in the 1950s and 1960s, claimed its critics, stifled the initiative of poor people and denied recipients the chance to gather resources to break out of their inertia and make independent and sustainable choices about how to live their lives. In this context, welfare reformers argued that welfare should be replaced with "workfare," a system that would require poor people within a given time to start on their own path to self-definition, self-actualization, and self-expression by searching for and securing a job of their choice, a solution to the problem that ruled out not only the participatory innovations of the 1950s and 1960s but also the creation of massive federal, state, or municipal work relief programs such as the U.S. Civilian Conservation Corps of the 1930s, the very name of which suggested in the late twentieth century a depersonalizing authoritarian discipline focused on a group of people with a single need defined by "outsiders" rather than the particular needs of particular self-defined individuals.

In this perspective, then, an antigovernment/individual empowerment animus ranks as a hallmark of national urban policy in the last half of the twentieth century, and few observers of the urban scene after 1950 advocated the renunciation of cultural individualism, of the right of individuals to self-definition, self-actualization, and self-expression. But the practice of cultural individualism did generate complaints about its excesses, including complaints about a rancorous incivility in the conduct of public life, examples of which included riots by African Americans and interracial violence in big cities; antiabortion street demonstrations, bombings, and shootings; an unwillingness to compromise in urban politics; and ad hominem attacks in political campaigns and in the conduct of everyday activities in politics and government.[38] Critics of incivility in political life pointed to several potential sources of it, including the spectacular decline of civility in big-city print and electronic media, especially in the so-called tabloid sensationalist press, and in big-city radio and television talk shows, arguably most spectacularly in the famously (or infamously) popular daytime personal/sexual relations maximum self-expression talk program hosted by Jerry Springer, a former Cincinnati city council member and mayor and for a time a thoughtful, respectable, witty, smart, discreet, and articulate television news anchor and commentator.

Worries about the excesses of cultural individualism also tended to focus on a concern for what critics regarded as a lack of "community" in American metropolitan areas generally, not just in the poorest and most cut-off neighborhoods, such as inner-city ones. But most of these critics did not assert that city dwellers seldom or never associated closely in the pursuit of shared interests of some sort. On the contrary. Later twentieth-century community grouses complained that, if anything, most city dwellers indulged too much in those kinds of communities, in communities of the like-minded.

The late-twentieth-century concern with community, that is, tended to focus on the absence of a sense of community that took as its common concern the welfare of the whole, a sense of civic community that placed the public interest on a par with individual and group interests. Advocates of promoting the idea of civic community as a supplement to psychological/cultural community sought not to impose on neighborhoods, cities, suburbs, or metropolitan areas their or anyone else's definition of the public interest. Instead, they sought to encourage individuals to develop a sense of empathy for, rather than a mere tolerance of, others and otherness. Such a sense of empathy, they argued, could provide a base for encouraging and facilitating the ability of individuals and groups to raise the basic issue of what kind of city they wanted themselves and others to live in some twenty or so years down the road. In debating that question, individuals and groups would have a chance to develop a yardstick against which to measure the relative importance of the claims of various groups, a yardstick useful in prioritizing those claims on the basis of compromise rather than in terms of the win-or-lose calculus so characteristic of fights over urban policy after 1950.[39]

But welfare of the whole public interest advocates attracted during the last half of the twentieth century little attention and exerted even less influence, a fact that projects bleak prospects for the future of American cities. Yet some grounds for hope can be found. These include the persistence of the golden rule as a common feature of all the world's great religions, each of which counts followers in American cities. And some have pointed to Václav Havel, who himself knows something about the nature of the human spirit, who has contended that the human capacity for self-transcendence rests at the root of all societies and runs, as he puts it, "infinitely deeper in human hearts and minds than political opinion, convictions, antipathies or sympathies."[40]

Yet few of those optimistic about the prospects for civic community and civic loyalty have pointed to another hopeful sign, the ascendancy in urban policy circles of baby boomers, people who caught on to and caught up with in the 1960s the rapidly spreading cultural individualistic trend, people commonly thought of still as self-centered cultural individualists par excellence. Some of these boomers, however, have begun to reevaluate their values in ways that point to a new and sometimes sentimental respect on their part for the importance of civic community and loyalty, for the public interest defined as the welfare of the whole social entity of concern.

A touching example of boomers' new interest in civic community and loyalty appeared in a recent reflection by Maureen Dowd, herself an aging boomer and a widely read columnist for the *New York Times*, on the killing in Washington, D.C., of two Capitol security guards by a deranged and armed intruder bent on shooting either members of Congress or tourists or both. Citing her own reaction to this tragedy, aging boomer President Bill Clinton's eulogy at a memorial service for the slain policemen, and aging boomer and moviemaker Steven Spielberg's just-released film about the D-Day invasion during World War II, all of which pointed to civic loyalty and sacrifice as an estimable quality, Dowd contended with a pinch of self-pity that baby boomers now realize "suddenly" that "we are going to die without experiencing the nobility that illuminated the lives of our parents and grandparents," who "lived through wars and depressions, life and death, good stomping evil." Instead, she added caustically, the boomers' "unifying experience was 'Seinfeld'" a fabulously popular 1990s television situation comedy about a small group of supremely self-centered and civically mindless friends.

Dowd concluded, however, on a dual note, one of regret and one of hope. "We live," as she put it, "in a culture more concerned with celebrity than morality, more centered on self than others." But at least "we're no longer demonizing the wrong people," cops and soldiers who by their career choices stand and fall as prime exemplars of civic concern, civic loyalty, and civic sacrifice.[41] And she might have gone further to promote optimism about the future of national urban policy by reminding us that even the most culturally individualistic of aging boomers carry within them a profound belief in the possibility of self-reinvention as well as the capacity for self-definition, self-actualization, and self-expression.

Yet we can't, perhaps, expect (or fear) too much, as the stance of two of the most formidable fin de siècle politicians suggest. Both President Bill Clinton, a "new Democrat," and prime (in 1999) GOP presidential candidate George W. Bush, a "compassionate conservative," like the authors of a widely read Progressive Policy Institute book called *Reinventing Government*, "believed deeply in government" and sought an activist role at the federal level in policymaking. But both of the politicians, like the book's authors, did not look back to the New Deal but adapted the key idea of national policy-makers between 1950 and the early 1970s by searching for new and politically and fiscally feasible ways of using federal authority to facilitate the pursuit of choices by individuals in their pursuit of self-identification and self-realization by designing a government that worked more like the private sector, and by selling the idea with market-oriented jargon.[42]

We cannot know as we enter the twenty-first century how these or other attempts to revive a patriotism centered on the public interest will turn out, though the effort seems worth a try. For we can affirm that the death of the idea of the city as a culture-generating force proved the most important single urban policy-making event of the last half of the twentieth century. It terminated the impulse to engineer the cultures of whole groups of people by the manipulation of the metropolitan social and physical environment. It also made possible the forging since the 1940s of two subperiods in the making of national urban policy. In the first, local, state, and federal governments aggressively sought to engage citizens and private sector institutions as partners in policy-making and implementation processes for the purpose of broadening lifestyle choices for everyone. In the second, the conviction that citizens and private sector institutions could do this more effectively on their own fueled an antigovernment impetus that touted deregulation, privatization, and grass-roots enterprise and empowerment with the minimum feasible participation of governments as the best routes to the same goal. These stunning transformations suggest that we may yet find culturally individualistic ways of making the welfare of the city as a whole once more our top urban policy priority, if we choose.

Notes

1. Thomas Lynch, "Socko Finish," *New York Times Magazine*, July 12, 1998, 36.
2. For longer introductions to federal urban policy in the periods covered in this chapter, see Kenneth Fox, *Metropolitan America: Urban Life and Urban Policy in the United States, 1940–1980* (Jackson: University Press of Mississippi, 1986); Carl Abbott, *Urban America in the Modern Age: 1920 to the Present* (Arlington Heights, Ill.: Harlan Davidson, 1987); and the admirably comprehensive list covering the Truman administration through the first Clinton administration in Robert J. Waste, *Independent Cities: Rethinking U.S. Urban Policy* (New York: Oxford University Press, 1998), 43–98. Waste helpfully covers subfederal governmental policies in chapter 5.

   Most of the argumentation and information in this essay have been documented in Zane L. Miller, *Suburb: Neighborhood and Community in Forest Park, Ohio, 1935–1976* (Knoxville: University of Tennessee Press, 1981); Zane L. Miller and Bruce Tucker, *Changing Plans for America's Inner Cities: Cincinnati's Over-the-Rhine and Twentieth-Century Urbanism* (Columbus: Ohio State University Press, 1989); and Robert B. Fairbanks, *For the City as a Whole: Planning, Politics, and the Public Interest in Dallas, Texas, 1900–1965* (Columbus: Ohio State University Press, 1998). I have tried to bolster that documentation in notes to this chapter. See also the review of post–1950 social science literature on urban policy in Roger Hansen, "Invitation to Annexation: Metropolitan Fragmentation and Community in Cincinnati and Houston, 1920–1980" (Ph.D. diss., University of Cincinnati, 1999), ch. 8; the treatment of literary figures James T. Farrell and Paul Goodman in Lewis S. Fried, *Makers of the City* (Amherst: University of Massachusetts Press, 1990), especially 146–148 on Farrell's conversion to cultural individualism; and the study of novelists and other writers about cities in the 1950s and 1960s, Carlo Rotella, *October Cities: The Redevelopment of Urban Literature* (Berkeley: University of California Press, 1998), which argues that these intellectuals found existing ways of thinking about cities inadequate for the 1950s and 1960s.

3. Demands for customization and specialization not only stymied efforts to create regional multifunctional governments but also sparked the proliferation of single-service districts, sometimes called authorities. Relatively rare in 1950, they numbered 33,131 in 1992, a figure 14,000 higher than the next most popular form of local government, the municipality. Kathryn A. Foster, "The Fish Stores of Government and Other Musings on Specialization," *Insight: The Journal of the School of Architecture and Planning, State University at Buffalo* 3 (1995): 48.
4. For a striking and characteristic example of the interplay of these themes in personal identity politics during the early 1970s, see Michael Novak, *The Rise of the Unmeltable Ethnics* (1971; New York: Macmillan, 1973), a mistitled and impassioned plea for the creation and mobilization of white ethnic group consciousness, and one that defined (37) the "youth culture" of the 1960s as essentially ethnic. The more sober concern among antideterministic social scientists for the ethnoreligious revival may be traced to the 1950s and 1960s as presented in Nathan Glazer and Daniel Patrick Moynihan, *Beyond the Melting Pot: The Negroes, Puerto Ricans, Jews, Italians and Irish of New York City* (Cambridge: MIT Press, 1963). The book presents ethnic identity not as a "given" but as something devised and chosen by groups of people as a strategy to cope with particular circumstances and as part of an American ethos of group formation on a nonnational and nonracial basis that started in the early nineteenth century (see vi and 291).

   Social scientists and social commentators differed in the 1960s and 1970s over whether white ethnicity would disappear as a factor in city life as more and more people chose a nonethnic way of life, but all of the most notable combatants agreed that both ways of life existed to one degree or another and presented a rich gamut of choices. See, for example, Harvey Cox, *The Secular City: A Celebration of Its Liberties and an Invitation to Its Disciplines* (New York: Macmillan, 1965); Daniel Callahan, ed., *The Secular City Debate* (New York: Macmillan, 1966); and Andrew M. Greeley, *Ethnicity in the United States: A Preliminary Reconnaissance* (New York: John Wiley and Sons, 1974), who defined ethnicity partially as "having to do with subtle but important differences of expectations in one's most intimate personal relationships" and suggested on the next page that intellectuals might comprise an ethnic group (31–32).
5. For a characteristic expression of the conscience of a Christian conservative, see the quotation from Donald E. Wildom in Bruce Borland and Jessica Bayne, eds., *American through the Eyes of Its People: Primary Sources in American History*, 2nd ed. (New York: Longman, 1997), 352–353. For a characteristic and influential apotheosis of the First Amendment and other freedoms of choice from a non-Christian conservative perspective, see Milton Friedman and Rose Friedman, *Free to Choose: A Personal Statement* (New York: Harcourt Brace Jovanovich, 1979), esp. 298–310.
6. John W. Whitehead, for example, the president of the "conservative" Rutherford Institute, recently described himself as a person of Marxian views in the early 1970s who later "went on to practice law, start a family, become a Christian, change some of my political views and establish an organization that specializes in civil liberties and religious rights." John W. Whitehead, "The Candidate," *New York Times*, September 16, 1998.
7. See Peter Steinfel's review of *American Evangelicanism: Embattled and Thriving*, by Christian Smith (Chicago: University of Chicago Press, 1998), in "Beliefs," *New York Times*, November 14, 1998. Steinfel helpfully points out that this view contradicts the socially deterministic notion that true believers flourish only when isolated from what Walter Lippmann called in 1929 "the acids of modernity."

8. Besides the social scientists and social critics cited in Miller, *Suburb*, and Miller and Tucker, *Changing Plans*, see also Alan Brinkley, "The Two World Wars and the Idea of the State," in *The Liberal Persuasion: Arthur Schlesinger, Jr., and the Challenge of the American Past*, ed. John Patrick Diggins (Princeton, N.J.: Princeton University Press, 1997), 131), who adds to the list of postwar antitotalitarians who warned in the 1950s about the excessive use of state power in democracies the names of Reinhold Niebuhr and Walter Lippmann. See also Richard Bernstein, "Foxes, Hedgehogs and the Defense of Freedom," *New York Times*, November 25, 1998, a review of Michael Ignatieff's biography of Isaiah Berlin, perhaps the most famous political philosopher of the last half century. An advocate of choice, Berlin also urged "the acceptance of the fact that tragedy inheres in choice, because there is no choice that leads to the solution of all problems." Berlin called for "less Messianic ardor, more enlightened skepticism, more tolerance of idiosyncrasies." As Ignatieff phrased Berlin's position, "It was individual freedom, to choose well or ill, which had to be defended, not some ultimate vision of the human good." Louis Wirth turned against the idea of the social determination of cultural groups as early as 1936 because he (and Karl Mannheim) by then regarded totalitarianism as a virtual inevitability in modern pluralistic societies. See Zane L. Miller, "Pluralism, Chicago School Style: Louis Wirth, the Ghetto, the City, and 'Integration,'" *Journal of Urban History* 18, no. 3 (May 1992): 262–264. The trend toward cultural individualism and away from deterministic views of the origin cultural groups seems also to have been a major event in Europe generally. See, for example, Tony Judt's review of *Dark Continent: Europe's Twentieth Century*, by Mark Mazower (New York: Alfred A. Knopf, 1999), in the *New York Times Book Review*, February 7, 1999, 11–12, and the book itself, esp. x–xi, where Mazower identifies the 1940s as the moment when the group-based ideologies of the 1910s, 1920s, and 1930s faded and "[p]eople rediscovered democracy's quiet virtue—the space it left for privacy, the individual and the family."
9. Serious people talked in the second quarter of the twentieth century and after 1950 about the establishment by the federal government of a coherent national urban policy. See, for example, U.S. National Resources Committee, Research Committee on Urbanism, *Our Cities: Their Role in the National Economy* (Washington, D.C.: Government Printing Office, 1937); Daniel P. Moynihan, ed., *Toward a National Urban Policy* (New York: Basic Books, 1970). Authors of the first study took seriously the possibility of forging such a policy, but Moynihan, the person selected by President Nixon to oversee this second effort, expressed deep reservations about the chances of reaching such a goal. Moynihan, like most of the authors he recruited for contributions to his study, deplored the absence among American city dwellers of "the essence of community," which Moynihan defined in a distinctively late-twentieth-century way as "the self-imposed decision to behave properly and as expected," a definition of the problem that, in the age of cultural individualism and hyperdiversity, rendered it virtually insoluble (*Toward a National Urban Policy*, 5). That dilemma prompted Moynihan to conclude his analysis by seeking divine assistance via an Irish prayer to Jesus to "take away our hearts o' stone, and give us hearts o' flesh. Take away this murderin' hate, and give us Thine own eternal love" (ibid., 334–336).
10. For Beard's view, see Charles A. Beard, "Written History as an Act of Faith," *American Historical Review* 39, no. 2 (January 1934): 219–229. The phrase "middle range theory" is most commonly associated with Robert K. Merton, one of the first and most prominent antideterministic sociologists. Merton noted regretfully in 1938 that "radical relativism" could be used dishonestly for "propagandistic purposes" by totalitarians to discredit advocates of middle range theory. Merton's quote may

be found in Peter Novick, *That Noble Dream: The 'Objectivity Question' and the American Historical Profession* (Cambridge: Cambridge University Press, 1988), 289, who reads it rather differently.

11. The assault on the social (and biological) determination of culture may be dated to the late 1930s. See, for example, Miller, "Pluralism," 251–279. Some have construed the attack on the idea of the social (and biological) determination of culture as a retreat from if not an attack on the idea of relativism. See, for example, Novick, *That Noble Dream*, chs. 10 and 11 and esp. 296–297. But it seems that the ostensibly "objectivist" social scientists and historians of the mid-twentieth century sought to hyperpluralize society, to explain somehow why so many individuals chose to escape from the prison of social or biological determinism and to behave in ways antithetical to their "given" class, or ethnic, or biological interests/ideology and in ways that sometimes sustained and sometimes undermined the efficient functioning of a particular political and/or economic system (such studies, of course, performed double duty as manuals of how either to sustain or undermine a particular system). In the process, most of them ended up in the camp that eschewed explanatory and therefore predictable "causes" (either rationally determined or as an "act of faith") for the "empirical" and "rational" measurement of trends, tendencies, and probabilities (correlations). They also acknowledged cultural individualism as a fact of American life and facilitated what I have called the "psychologizing of American life" (in the past, present, and future) by psychologizing the analysis of American life. See, for example, Robert K. Merton, *Social Theory and Social Function: Toward the Codification of Theory and Research* (Glencoe, Ill.: Free Press, 1949); Theodor W. Adorno et al., *The Authoritarian Personality* (New York: Norton, 1950); Daniel Bell, ed., *The Radical Right: The New American Right Expanded and Updated* (1955; Garden City, N.Y.: Doubleday, 1963); Seymour Martin Lipset, *Political Man: the Social Bases of Politics* (1960; Garden City, N.Y.: Doubleday, 1963); Samuel J. Eldersveld, *Political Parties: A Behavioral Analysis* (Chicago: Rand McNally, 1964), esp. v, 1–23, "A Theory of the Political Party," an idea developed in the early and mid–1950s that virtually pluralized out of existence the party as a describable operational mechanism.

    Merton's idea of "latent" functions ranks among his most influential contribution to urban policy-making. It normalized and "centered" lifestyle choices (such as crime and urban political bossism) previously regarded as aberrant and destabilizing because they helped sustain in the United States democratic capitalism (a notion applicable, of course, to other systems of political economy) as the dominant value system. Merton saw people formerly regarded as "dysfunctional" and therefore as potential "dependents," or "social problems," or revolutionaries, as quite capable of making choices from among "values" exalted by the democratic capitalistic system (hard work, enterprise, competition, cooperation, and loyalty, for example) and using them to forge lifestyles and profiting in ways that made them independent of rather than dependent on "legitimate" institutions and supporters of the status quo instead of revolutionaries. The moral: people opposed to crime or bossism should broaden the lifestyle choices available to criminals and bosses and their clients, either by improving or transforming the dominant value system. For a useful but neglected account of Merton in the late 1930s and 1940s as a self-consciously antideterministic sociologist and as an antifascist defender of science and democracy, see David A. Hollinger, "The Defense of Democracy and Robert K. Merton's Formulation of the Scientific Ethos," *Knowledge and Society: Studies in the Sociology of Culture* 4 (1983): esp. 4, 10–12. Merton sought in these years to systematize and codify various strands of functional analysis, an effort that may be seen as a prerequisite for shifting it from a deterministic to a culturally individualistic base.

12. The psychologizing of American society by the 1990s yielded a new tendency, the labeling of a growing list of behavioral problems, including sexual adventureness, as psychological addictions amenable to clinical therapy. See Joe Sharkey, "'Enabling' Is Now a Political Disease," *New York Times*, September 27, 1998. This addiction phenomenon, like other aspects of the psychologizing of American society, helped strew post–1950 metropolitan terrains, especially suburban ones, with unprecedentedly large numbers of psychologists' offices and psychology clinics and metropolitan daily newspapers and radio and television talk shows with an army of psychology advice-givers, subjects largely neglected by urbanistic scholars.
13. These turf wars sometimes took on multicultural dimensions. In Atlanta's Kirkwood, a predominantly African American neighborhood, more affluent white newcomers, many of them gay, sought in the late 1990s to assist in driving out drug dealers by staging a series of vigils. The antidrug protesters subsequently came under fire not only from drug dealers but also from "clean" African Americans who feared that the white "gentrification" of their neighborhood would lift real estate values and drive blacks out of the neighborhood. In the vigils and counterprotests, some African Americans moved beyond the "class" issue and expressed both anti-white and antigay sentiments. See Lauren Keating, "Growing Pains: Tensions Break Out in Kirkwood Neighborhood," *Creative Loafing*, June 20, 1998, 22. Writers and journalists critical of the metropolitan daily newspapers in every big city produced after the 1950s "alternative" sheets, such as *Creating Loafing*, usually weeklies, which focused on popular entertainment and "little" local stories in an effort to establish for themselves a niche in the communications market among younger people and dating singles, both hetero- and homosexual.
14. The *New York Times* sniffily characterized the southern captivity of Saks as "no big bargain" because the purchaser's stock, Profitt, Inc., dropped by $4 the day after announcement of the deal. *New York Times*, July 7, 1998. A week later Little Rock–based Dillard's Inc. signed a letter of intent with Charlotte, North Carolina–based Belk Stores Services Inc. to swap seven stores of Fairfield, Ohio–based Mercantile Stores Co., Inc. for nine Belk stores. *Cincinnati Enquirer*, July 15, 1998.
15. The *New York Times* noted in 1998 that state and local governments since the 1950s had competed for business corporations by offering expensive incentives such as "free factories, canceled property taxes, low-interest loans, free access roads, and special tax credits." In the 1990s twenty-one localities, most of them in the South and Midwest, came up with yet another carrot, government paybacks for periods of up to ten years of a percentage of the wages paid to workers in new businesses (4.5 cents on every dollar paid, for example, to workers in a Tulsa, Oklahoma, stove factory). Louis Uchitelle, "Taxes Help Foot the Payrolls as States Vie for Employers," *New York Times*, August 11, 1998.
16. Humphrey's short but powerful speech is reprinted in Paula Wilson, ed., *The Civil Rights Rhetoric of Hubert H. Humphrey, 1948–1964* (Lanham, Md.: University Press of America, 1984), 3–5. Antideterministic views such as Humphrey's on the race question found distinguished contemporary social science support in Gunnar Myrdal's classic *An American Dilemma* (1944; New York: McGraw-Hill, 1964), 2 vols. See especially vol. 1, the preface, introduction, and 520. Myrdal's study provides a long but nifty example of how "objectionist" middle range theorists got around the politically paralyzing implications of the relativistic principle of uncertainty and/or unknowability. The key rested in discovering and accepting someone else's "values" as a representation of what reality should be (what he called "the American Creed") as the basis for deciding what might and should be done next. This could be done both discriminately and pluralistically, as Isaiah Berlin has explained, by recognizing all value systems, including Nazism, as equally hu-

man and equally understandable but not therefore necessarily as equally tolerable. For a fuller yet concise explication, see Isaiah Berlin, "My Intellectual Path," *New York Review of Books*, May 14, 1998, esp. 58–59.

17. Arthur M. Schlesinger Jr., *The Vital Center: The Politics of Freedom* (Boston: Houghton Mifflin, 1949), ix. This easing may also be seen as a product of the revolt against the idea of the social determination of culture. See, for example, Roger Daniels, "Immigration since World War II: The Need for a New Paradigm," *Polish American Studies* 55, no. 1 (spring 1998): 41.
18. The city of Baltimore tried out in the early 1950s the idea of urban renewal as the rehabilitation of an entire neighborhood using citizen-participation techniques, a notion that became conventional wisdom in a few years, namely that slum dwellers could be taught in short order to take control of their lives and make residential as well as other lifestyle choices. See Miles L. Colean, ed., *The Human Side of Urban Renewal: A Study of the Attitude Changes Produced by Neighborhood Rehabilitation* (Baltimore: Fight-Blight, 1958), esp. vii–viii.
19. Waving the bloody flag of totalitarianism in the campaign for the devolution of government authority persisted in respected quarters well after the initial antideterministic impulse of the 1940s and 1950s. Nobel Prize–winning economist Milton Friedman and his wife, Rose, for example, claimed in 1979 that the "tide toward . . . New Deal liberalism" had crested in the 1970s, but they found "as yet no clear evidence whether the tide that succeeds it will be in the spirit of [Adam] Smith and Jefferson or toward an omnipotent monolithic government in the spirit of Marx and Mao." Friedman and Friedman, *Free to Choose*, 284–285. The devolutionary Friedmans also claimed that special economic interests swamped the "general interest" in large cities, states, and Washington, D.C., and cited villages, towns, small cities, and suburban areas as the only places where one could find "anything like effective detailed control of government by the public . . . , and even there only to those matters not mandated by state or federal governments" (295). For a neo-Marxian proponent of promoting the freedom to choose as a national urban policy, see Michael P. Smith, *The City and Social Theory* (New York: St. Martin's Press, 1979), esp. 268.
20. Maximum feasible participation policy intended to engage all citizens, not just the poor, in solving urban problems and, as we shall see, could be defined both to involve and not to involve extensive government funding, organization, and supervision of the participatory process. In both cases, however, participation may be seen as a therapeutic process to assure "the ultimate integrity of the individual," for in both cases it encouraged citizens to seek self-fulfillment through self-actualization in a process that taught them how to define and realize their own lifestyle destiny, a culturally individualistic learning-by-doing exercise. On the origins in the 1950s of maximum feasible participation through community action, see Zane L. Miller and Bruce Tucker, "The Revolt against Cultural Determinism and the Meaning of Community Action: A View from Cincinnati," *Prospects: An Annual of American Cultural Studies*, vol. 15 (Cambridge: Cambridge University Press, 1990), 413, 435 n.2; and Daniel Patrick Moynihan, *Maximum Feasible Misunderstanding: Community Action in the War on Poverty* (New York: Free Press, 1969), ch. 1, esp. 8–19.
21. On the suburban mixing of white ethnics after 1950, see Jerome Weidman, *The Enemy Camp* (New York: Random House, 1958); and Eric Monkonnen, "Community on the Edge," *The Long View: A Journal of Informed Opinion* 4, no. 2 (spring 1998): 31–32. On the political effects of that phenomenon, see Walter Johnson, *1600 Pennsylvania Avenue: Presidents and the People, 1929–1959* (Boston: Little, Brown, 1960), 267–269; Milton Rakove, *Changing Patterns of Suburban Politics:*

Cook County, Illinois (Chicago: Center for Research in Urban Government, Loyola University-Chicago, 1965); and Zane L. Miller and Patricia Mooney Melvin, *The Urbanization of Modern America: A Brief History* (San Diego: Harcourt Brace Jovanovich, 1987), 216–218. For a proposal to use government for the purpose of promoting racial and class mixing in outlying areas, see Anthony Downs, *Opening Up the Suburbs: An Urban Strategy for America* (New Haven, Conn.: Yale University Press, 1983). For an earlier take on the same idea, see Charles Abrams, *Forbidden Neighbors: A Study of Prejudice in Housing* (New York: Harper and Brothers, 1955), a scathingly prochoice attack on residential segregation by class, religion, ethnicity, and race and on social deterministic ideas that undergirded the practice. The notes document and date the assault on this new problem to the late 1940s, and Abrams himself moved toward the idea of providing blacks a choice between residential segregation and integration in 1946, in Charles Abrams, *The Future of Housing* (New York: Harper and Brothers, 1946), 404–405. For the advocacy in the 1960s of increased residential choices for African American urbanites, see *Report of the National Advisory Committee on Civil Disorders* (New York: Bantam Books, 1968), 395–408, which stressed choices for policy-makers as well as for African Americans seeking a place to live. For two of the most influential books among urban planners and policy-makers after 1960, see Jane Jacobs, *The Death and Life of Great American Cities* (New York: Random House, 1961); and Herbert J. Gans, *The Urban Villagers: Group and Class in the Life of Italian-Americans* (Glencoe: Free Press, 1962). Both Jacobs and Gans assaulted the idea of slums as social and physical environments that determined the identities and behavior of their residents. See, for example, Gans, *The Urban Villagers*, 308–317. For the remarkable durability of Jacobs's approach, first published in 1957 as "Downtown Is For People" in *The Exploding Metropolis*, ed. William H. Whyte Jr. (New York: Time, 1957), 157–184, see Roberta Brandes Gratz, *The Living City* (New York: Simon and Schuster, 1989); and Roberta Brandes Gratz with Norman Mintz, *Cities Back from the Edge: New Life for Downtown* (New York: John Wiley and Sons, 1998).

22. Changes in the federal accelerated depreciation tax break on commercial buildings sparked a huge suburban building boom in the 1980s, just as the invention of that device ignited the proliferation of shopping malls in the 1950s. See Tom Hanchett, "Talking Shopping Center: Federal Tax Policy, Commercial Sprawl, and the Decline of Community," *The Long View: A Journal of Informed Opinion* 4, no. 2 (spring 1998): 47–50. For a culturally individualistic celebration of the edge city phenomenon that traces its origins as an ideal urban form to Frank Lloyd Wright, *The Living City* (New York: Horizon Press, 1958), see Joel Garreau, *Edge City: Life on the New Frontier* (New York: Doubleday, 1988), esp. 10–11.
23. Clerk of Council of the City of Cincinnati, Zoning Code for the City of Cininnati, January 1983, typescript, loose-leaf notebook. Pages 437–438 of this document cite the date of each change.
24. The first step to enclosed stores in malls seems to have taken place in the 1970s, when wall-less malls and downtown department stores began to refer to particular open-floor spaces as occupied by "boutiques," localities defined by dictionaries as specialty shops.
25. Foster, "The Fish Stores," 51, 54 n.13 and n.14. Foster concludes her piece with warnings about the shortcomings of specialization but adds, ironically, a quotation from science fiction writer Robert Heinlein celebrating the individual's capacity for specializing in multiple identities. Heinlein said, "A person should be able to: heal a wound, plan an expedition, order from a French restaurant, tell a joke, laugh at a joke, laugh at oneself, cooperate, act alone, sing a children's song, solve equa-

tions, throw a dog a stick, pitch manure, program a computer, cook a tasty meal, love heartily, fight efficiently, die gallantly. Specialization is for insects." Ibid., 53.

26. Ibid., 54 n.13.
27. Paul S. Boyer, *Promises to Keep: The United State since World War II* (Lexington, Mass.: D. C. Heath, 1995), 471–474.
28. These other examples include the establishment by black developers of very expensive suburban communities exclusively for wealthy blacks. See Tim O'Reiley, "Black Atlanta Builders Aim at High End," *New York Times*, November 22, 1998. As one appreciative resident put it, "Thirty years ago, twenty years ago, we probably would have had to move into a majority white neighborhood to get the kind of home we wanted. Now, it is not necessary to move in with white families to say that we've made it."
29. For a different take on the 1970s as a period of changing directions, see Nicholas Lemann, "How the Seventies Changed America," *American Heritage* 42 (July/August 1991): 39–42, 44, 46, 48–49, reprinted in Stephen B. Oates, *Portrait of America, vol. 2* (Boston: Houghton Mifflin, 1995), 438–448. Though well beyond the scope of this chapter, one might argue that American social science literature, like American historical literature, offered after 1950 essentially two tendencies in explaining the general stance of individuals in the establishment of their identities and lifestyles. From the 1950s into the 1970s social scientists tended to present individuals as responding to changing social realities by choosing from a range of options determined by those social realities. From the 1970s through the 1990s social scientists tended to present individuals as constructing, deconstructing, and reconstructing social reality by their attempts to define and establish their identities and lifestyles. For studies that may be seen as supporting this view, see Novick, *That Noble Dream*, esp. 522–629; William B. Hixson Jr., *Search for the American Right Wing: An Analysis of the Social Science Record, 1955–1987* (Princeton, N.J.: Princeton University Press, 1992), esp. xxv–xxvii, 289–292, 307–327; and Terrence J. McDonald, ed., *The Historic Turn in the Social Sciences* (Ann Arbor: University of Michigan Press, 1996), esp. 1–14. The policy implications of the two approaches differ, of course, and help explain the popularity of the term "culture wars" from the 1970s to the present and governmental efforts in the 1950s and 1960s to alter the social and physical environment for the purpose of widening the range of lifestyle choices and of liberating groups seen as victimized by social and/or physical conditions.
30. Boyer, *Promises to Keep*, 447.
31. Daniel Patrick Moynihan used the phrase "benign neglect" with respect to the issue of race in a memo to President Nixon dated March 2, 1970. According to William Safire, this term in its memo context meant that extremists on both sides of the question should lower their voices, but some misinterpreted it to mean that the administration's policy with respect to race should be one of benign neglect. But that "misinterpretation" is certainly understandable in the light of Safire's account of where Moynihan got the phrase. It came, says Safire, from the earl of Dunham, who told Queen Victoria in 1839 that Canada had done so well during a period of benign neglect by the British government that Canadians should be granted self-rule. William Safire, *Safire's New Political Dictionary* (New York: Random House, 1993), 49. And the "misinterpretation" comported well with my reading of Moynihan's views on race published in the 1960s.
32. On the family proposal and the sense by 1966 of the political infeasibility of effective governmental action on race, see Lee Rainwater and William L. Yancy, *The Moynihan Report and the Politics of Controversy* (Cambridge: MIT Press, 1967), esp. 480–482. The idea of no-strings payments to the poor came from Chicago

economist Milton Friedman. See George Mowry and Blaine A. Brownell, *The Urban Nation, 1920–1980* (New York: Hill and Wang, 1981), 200. On the immigrant analogy and its shortcomings, see Richard C. Wade, "Historical Analogies and Public Policy: The Black and Immigrant Experience in Urban America," in Diggins, *The Liberal Persuasion*, 185–196. On the economic shortcomings of the family strategy, see Herbert G. Gutman, *The Black Family in Slavery and Freedom, 1750–1925* (New York: Pantheon Books, 1976), esp. 461–469. Scatter-site proponents posited the persistence of involuntary racial residential segregation as the fundamental obstacle to the liberation of African Americans and to the undermining of racism among whites. See, for example, Wade, "Historical Analogies"; Kenneth B. Clark, *Dark Ghetto: Dilemmas of Social Power* (New York: Harper and Row, 1965); and Morris Milgram, *Good Neighborhood: The Challenge of Open Housing* (New York: Norton, 1977), the last of which not only consists of a plea for racial residential integration but also an unusually optimistic how-to-do-it manual by a man who launched his integrationist real estate business in 1952 in Philadelphia. But Moynihan had advocated in 1963 benign neglect as the proper national policy for African Americans and saw the shoring up of the allegedly pathological black family as a way of assisting middle-class blacks in uplifting lower-class blacks without significant additional governmental intervention. See Glazer and Moynihan, *Beyond the Melting Pot*, 196. Another prominent "benign neglecter" of the 1960s took this idea several steps further by arguing in 1968 that government should not try to solve any urban social problem on the grounds that an urban crisis did not exist, that urban social problems would eventually solve themselves, and that government social programs would sustain such problems or make them worse. See Edward C. Banfield, *The Unheavenly City: The Nature and Future of Our Urban Crisis* (1968; Boston: Little, Brown, 1970), esp. 3–4, 10–11, 14, 21–22, 255–263. Moynihan did not go that far. He argued in 1969 that "the role of social science lies not in the formulation of social policy, but in the measurement of its results." Moynihan, *Maximum Feasible Misunderstanding*, 193. By 1969 Moynihan had held social science advisory positions in both Lyndon Johnson's and Richard Nixon's administrations, and he next took his own advice of 1969 on the role of social science by serving as a U.S. senator from New York, which enabled him to assist in the making of policy as a citizen and to evaluate it in his other identity as a social scientist.

33. Supply-side economics became associated primarily with the name of economist Arthur Laffer. For an introduction to and defenses of supply-side economics, along with critical essays on the subject at the onset of the first Reagan administration, see Richard H. Fink, ed., *Supply-Side Economics: A Critical Appraisal* (Frederick, Md.: University Publications of America, 1982). A pro-supply-side essay in this book, one of "the finest expositions in favor of supply-side economics . . . yet written," says Fink, contends that "the level of aggregation employed in Keynesian economics ignores the critical role of individual decision making . . . as regards the decisions to work, save, and invest" (xiv). For a neo-Marxian (that is, culturally individualistic Marxian) take on supply-side economics, see Robert L. Heilbroner's essay in Fink, *Supply-Side Economics*, 80–92, esp. 89–91. Here Heilbroner notes that certain elements of the bourgeois social order exercise a certain autonomy as promoters of political, social, and intellectual liberty; cheerfully concedes to "social orders [past, present, and presumably future] the right to commit suicide"; and argues (as an act of faith) the probability that present trends will continue over the next decade while not so cheerfully concluding that no one possesses a grand theory capable of predicting what might happen in the long run. His policy advice? Be wary of those who "wish to tilt the balance of power" in favor of businesspeople,

agents directly "connected with the system of wage labor on which the capitalistic economic systems rest."

34. Some commentators defined the growth strategies of the 1980s and 1990s as an economic phenomenon driven by "greed." But greed, like the poor, has always been with us and may be usefully interpreted according to its source as defined by people in different time periods. In this view, then, the late-twentieth-century version of greed may be seen not merely as a moral failing or an economic motive but as a psychological phenomenon, a form of behavior adopted by choice in the pursuit of self-fulfillment through self-definition, self-actualization, and self-expression. This perspective also helps explain the apparent paradox of raging corporate mergers in an age of individualism because it reminds us that corporate mergers are not new but that, in the final analysis, its practice in the last half of the twentieth century stemmed not from economic motives but from cultural individualism in a psychologizing society. American corporations adjusted with special facility to this new kind of individualism because they ranked under American constitutional law as individual persons possessed of all the rights and privileges of individual human citizens.
35. Social scientific debates after the 1960s over the relative role of genetics and family nurture as major sources of "outside" influence on educational achievement in public schools deflected attention from the effects of racially and economically segregated city neighborhoods on student behavior and attitudes toward learning. See, for example, Carol Tavris, "Peer Pressure," *New York Times Book Review*, September 13, 1998, 14–15. Tavris emphasized that nonconformist teenagers represented exceptions who proved the normal tendency, namely, that the vast majority chose to conform not as a matter of mere mimicry but as a strategy for survival, security, success, and status among their peers and that they tended to behave differently within their families (and, of course, after they grew up and selected various identities as adults).
36. Bob Herbert, "Renovating HUD," *New York Times*, October 18, 1998. One of these inner-city projects (in Atlanta) involved the renovation of a golf club built in the 1920s as the centerpiece of the new community, a common suburban practice after 1950. Clifton Brown, "Golf Course at the Center of a Community," *New York Times*, October 29, 1998. Public housing authorities borrowed this development style from the so-called new urbanists, critics of the narrow range of neighborhood choices in the suburbs and of their automobile centeredness which, these critics argued, suppressed neighborliness even in people who wanted a neighborly lifestyle. See Sarah Boxer, "A Remedy for the Rootlessness of Modern Suburban Life?" *New York Times*, August 1, 1998. See also the interesting set of essays on the question of whether "Americans have lost their sense of community and become more isolated" since 1950 in *The Long View: A Journal of Informed Opinion* 4, no. 2 (spring 1998), published by the Massachusetts School of Law at Andover. The issue consisted of ten essays, and the "ayes" carried the day, nine to one. All nine "ayes" also regarded the phenomenon as a bad thing. Celebration, Florida, a product of the Walt Disney Company, comprised the most famous and one of the most closely regulated of the new urbanistic community-building towns and did indeed, according to one resident, encourage neighborliness. See Douglas Frantz, "Living in a Disney Town, with Big Brother at Bay," *New York Times*, October 4, 1998.
37. In thinking about the definition of "welfare" in the 1950s and 1960s, consult *Webster's Seventh New Collegiate Dictionary* (1971). It defined "welfare" as the state of well-being, especially in relation to good fortune, happiness, or prosperity. It defined the "welfare state" as a social system based upon the assumption by a political state of primary responsibility for the individual and social welfare of its

citizens. These sound compatible with my view of the new directions in welfare practice during the 1950s and 1960s.

38. See, for example, Charles Mahtesian, "The Politics of Ugliness," *Governing: The Magazine of States and Localities* (June 1997): 18–22. The cover of that issue of *Governing* asked a question: "Why Can't They Behave? The Collapse of Civility in Local Government."
39. Civic community advocates who disagreed on a variety of particular issues agreed on this general one. See, for example, the essays in Moynihan, *Toward a National Urban Policy*; Friedman and Friedman, *Free to Choose*, ch. 10; Hadley Arkes, *The Philosopher in the City: The Moral Dimensions of Urban Politics* (Princeton, N.J.: Princeton University Press, 1981), esp. 3; Robert N. Bellah, *Habits of the Heart: Individualism and Commitment in American Life* (Berkeley: University of California Press, 1985), esp. 285–290, 307, lines 16–22; Miller and Tucker, *Changing Plans*, esp. 168; James Fallows, "How Journalism Undermines Public Life," *The Long View: A Journal of Informed Opinion* 4, no. 2 (spring 1998): 56–64; Jay Rosen, "Public Journalism: An Invitation to Join the Drama of Democracy," *The Long View: A Journal of Informed Opinion* 4, no. 2 (spring 1998): 65–72; David Rusk, *Cities without Suburbs* (Washington, D.C.: Woodrow Wilson Center Press, 1985), esp. 125–126, 129–130; and Thomas L. Haskell, "The New Aristocracy," *New York Review of Books*, December 4, 1997, 47–49. Worries about the absence of interest in civic duty led big-city school systems to require, or to consider requiring, teenagers to volunteer up to forty hours per year on community service projects. For some of the quandaries generated by this idea, see Dirk Johnson, "School Work, Homework and Good Works," *New York Times*, September 1, 1998. The article identifies some of the specific volunteer activities but does not, characteristically, talk about the question of how their organizers thought such service related to or advanced the public welfare defined as the welfare of the whole.
40. Quoted in Miller and Tucker, *Changing Plans*, 166.
41. Maureen Dowd, "An American Tragedy," *New York Times*, July 29, 1998. In *The Greatest Generation* (New York: Random House, 1998), Tom Brokaw, a television journalist and another graying boomer, also expressed nostalgic reverence for those who fought the Great Depression and World War II. For a much more optimistic boomer prediction that aging boomers will transform American society and cities for the better, see Theodore Roszak, *America the Wise: The Longevity Revolution and the True Wealth of Nations* (Boston: Houghton Mifflin, 1998). For Roszak, the fact of boomers living decades longer than their parents and grandparents has created a bank of citizens who could "join with others in building a compassionate society where people can think deep thoughts, create beauty, study nature, teach the young, worship what they hold sacred, and care for one another" because of their liberation by the longevity revolution from having "to worry about raising a family, pleasing a boss, or earning more money" (8). Roszak's forecast struck some observers as a politically naive leap of faith. See, for example, the review of Roszak's book by Mark Levinson (a labor union economist), "The New People," *New York Times Book Review*, October 4, 1998, 32. Other advocates of reviving the idea of the public interest saw hope in the growing bipartisan interest during the late 1990s in suburban precincts and statehouses for regulations to limit the sprawling spread into the countryside of shopping strips, residential subdivisions, and suburban office parks. Neal Pierce, "Issue of Sprawl Control Comes of Age with 1998 Elections," *Cincinnati Post*, November 27, 1998; and Todd S. Purdum, "Suburban 'Sprawl' Takes Its Place on the Political Landscape," *New York Times*, February 6, 1999.

42. David Osborne and Ted Gaebler, *Reinventing Government: How the Entrepreneurial Spirit Is Transforming the Public Sector* (Reading, Mass.: Addison-Wesley, 1992). Some limited-government Republicans accused the authors of arguing that the federal government should define "common goals," apparently unaware of the possibility and popularity of the use of public opinion polls to test programs *before* their proposal. See Edward H. Crane, "The Clintonesque George Bush," *New York Times*, August 4, 1999.

Chapter 9

The End of Liberalism

Narrating Welfare's Decline, from the Moynihan Report *(1965) to the Personal Responsibility and Work Opportunity Act (1996)*

WILLIAM GRAEBNER

BETWEEN 1965 and the end of the century, welfare—that is, Aid to Families with Dependent Children (AFDC)—expanded dramatically; came under attack from conservatives, libertarians, and liberals; and then, in the 1990s, was virtually eliminated as a federal program through legislation that had broad, bipartisan support. Throughout that process of growth and declension, social scientists played central roles in shaping perceptions of welfare, most significantly by examining the impact of welfare on the work ethic, on family structure, on gender relations, on poverty, and on inner-city black communities. This is an enormously complex story, and I have engaged it by focusing on four influential texts, each by a prominent social scientist: Daniel P. Moynihan's *The Negro Family: The Case for National Action* (1965), otherwise known as the Moynihan *Report*; Charles Murray's *Losing Ground: American Social Policy, 1950–1980* (1982); Martin Anderson's *Welfare: The Political Economy of Welfare Reform in the United States* (1978); and David T. Ellwood's *Poor Support: Poverty in the American Family* (1988). Although this approach inevitably oversimplifies somewhat, it also makes possible a more intensive critical reading of these key historical documents.

The documents tell the story of the decline of liberalism, tracked and facilitated by the social sciences. The story of the decline begins with the psychological, therapeutic, and historical perspective of the Moynihan

Report—liberal but, even then, cautious. It continues in the late 1970s and early 1980s, when psychological and historical understandings of the welfare experience and of human nature give way in the work of Anderson and Murray to a very different set of understandings, based on economics, the marketplace, and the transcendent value of work. Its concluding chapter takes place in the midst of the Reagan presidency, when a troubled and confused liberalism, represented here in the work of Ellwood, abandons its faith.

The Politics of the Therapeutic

The Negro Family (1965) appeared just as the civil rights movement was negotiating the turn to "black power." Not only was sociologist and Assistant Secretary of Labor Daniel P. Moynihan aware of the importance of that moment, but his *Report* was very much a response to its exigencies. Moynihan feared a revolution based on rising expectations, as blacks turned from the quest for legal equality and equal opportunity to another quest, more threatening to the society: this one for equality of results as a group. Moynihan did not say that this goal was wrongly conceived (although he did argue that it was different from, and essentially beyond, the traditional "white" understanding of equality); indeed, he seemed forthright in declaring that the nation must invest itself in the achievement of that goal. "The principal challenge of the next phase of the Negro revolution," he wrote, "is to make certain that equality of results will now follow." Moynihan's next words reflected the deep anxiety that underlay the *Report*. "If we do not," he wrote, "there will be no social peace in the United States for generations."[1]

Given the dire consequences that Moynihan predicted would follow from a failure to produce "equality of results," it is remarkable how little the *Report* is concerned with results of any kind. Most of the document is, of course, an indictment of the "crumbling," disorganized, dysfunctional, and pathological Negro family.[2] Although the title calls for "national action," none is advocated. The *Report* argues that because the Negro family is disorganized, blacks will inevitably be frustrated in their search for equality, implying that the black family must be repaired and brought back to health before progress toward the goal of equality of results can be made. On that level, the Moynihan *Report* seems to be less a road map to black equality than a roadblock. While ostensibly committed to the most far-reaching social goals, the *Report*'s most notable achievement is to conceptualize the black family as a substantial—perhaps insuperable—obstacle to any such plans. If only for that argument, the *Report* should be understood as a neoconservative document.

At the time of its release, the Moynihan *Report* was understood as a strikingly original document; its critique of the black family seemed to come out of nowhere to dramatically reshape and reconceptualize ongoing debates on welfare, the welfare state, civil rights, the inner cities, and the place of blacks in American life. But in many ways the *Report* looked not to the future but to the past for its inspiration, argumentation, logic, and ideology. As scholars have noted, Moynihan's argument that the black family is characterized by a "tangle of pathology" relies in part on Stanley Elkins's *Slavery* (1959), quoting extensively from Nathan Glazer's introduction to the 1963 edition of Elkins's book and, using a summary of the book by Thomas F. Pettigrew, drawing out Elkins's parallel between slavery and the Nazi concentration camps. Employing the Elkins parallel, Moynihan emphasized the extent to which slavery had damaged the black psyche and produced an obedient black male with a low need for achievement.[3]

A later generation of historians of slavery, including Herbert Gutman, Lawrence Levine, and John W. Blassingame, would reveal how flawed Elkins's perspective (and Moynihan's gloss on it) was, but in the early 1960s the dominant view in the social sciences was one of powerful institutions capable of crushing and defeating their passive and helpless inhabitants.[4] This literature owed much to growing concern about corporate bureaucracy in the period, but of equal importance was the Holocaust, which by the end of the 1950s had become, in Christopher Lasch's words, a "convenient symbol for the prevailing sense of helplessness."[5] Bruno Bettelheim staked out the territory with *The Informed Heart* (1960), which drew from the concentration camps the lesson that individuality was endangered by the standard features of mass society.[6] Perhaps the most important contribution to the genre was social psychologist Erving Goffman's *Asylums* (1961), an analysis and description of life in closed, "total" institutions, including prisons, monasteries, boarding schools, and mental hospitals. Moynihan's *Report* did not employ the concept of the total institution—perhaps because Goffman's account, for all its focus on institutional power, also suggested the possibilities and strategies of inmate resistance—but it otherwise shared Goffman's central concern with the survival of the self.[7]

The Moynihan *Report* contained a healthy dose of what historian Daryl Michael Scott has recently labeled "damage imagery." Blacks had been damaged not just by slavery but by what the *Report* called "three centuries of injustice," a phrase that seemed to imply that black problems were so deep and of such long duration that their solution would require Herculean efforts.[8] Much of the damage was to black men. The Moynihan *Report* claims that segregation was especially hard on the African American male; Jim Crow "worked against the emergence of a strong father figure." In a striking passage revealing ignorance of the black tradition of the "dozens," the *Report* asserts that the social requirements of a profoundly racist society had prevented blacks from learning and exhibiting the bold attitudes and postures required of successful men. "The very essence of the male animal," states the *Report*, "from the bantam rooster to the four-star general, is to strut. Indeed, in 19th century America, a particular type of exaggerated male boastfulness became almost a national style. Not for the Negro male. The 'sassy nigger' was lynched."[9] As a consequence, African American men had never learned to be men—had never learned to fulfill their responsibilities to provide food for young African American women. There was more damage to come. Migration to northern cities disrupted familial relations and dissolved family bonds, producing "family pathology." And, in the mid-twentieth century, thirty-five years of "disaster levels" of African American male unemployment had driven African American women into the workplace, undermining the "position of the father" and contributing to the community's harmful "matriarchal structure."[10] In a section of the *Report* dealing with juvenile delinquency, Moynihan emphasized the "personality effects" of a fatherless home environment.[11]

The Moynihan *Report*'s characterization of black society as pathological was important but not new. Swedish economist Gunnar Myrdal had used the word in *An American Dilemma* (1944), written for the Carnegie Corporation. Myrdal's influential book described the nation's racial problem as a "moral dilemma" within the heart of every white American—the same "racism" to which the Moynihan *Report* briefly referred. But Myrdal also suggested that any reasonable solution of the race problem would necessarily mean the assimilation of the inferior black culture, which *An American Dilemma* labeled a "pathology." "It seems frankly incredible," he wrote, "that the Negro people in America should feel inclined to develop any particular race pride at all."[12] Scott's *Contempt and Pity* (1997) reveals that among those invested in the idea of black pathology were sociologists E. Franklin Frazier and Kenneth Clark; novelists Richard Wright, Ralph Ellison, James Baldwin, and William Styron; black leaders Whitney Young and Martin Luther King Jr. (in his "Letter from Birmingham Jail"); historian Allen Matusow; and social anthropologist Oscar Lewis, whose "culture of poverty" approach was, Scott suggests, linked to ideas of black pathology. According to Scott, the pathology argument remained peripheral until 1954, when thirty-five social scientists, ignoring the absence of scholarly evidence, signed a statement attesting to the negative impact of segregation on the black psyche. Appended to the plaintiffs' brief in *Brown v. Board of Education*, the statement, and its apparent acceptance by the Warren Court, elevated the importance of the pathology argument within the civil rights movement.[13] Scott argues that damage imagery appealed to liberals and conservatives. The former found damage imagery useful in persuading the white middle class that segregation was harmful; the latter—the racists among them, anyway—could indulge in the image of a troubled black inferiority.

The term "pathology" is a medical one; it connotes disease, and its use implies the appropriateness of therapeutic intervention in order to return the patient to a healthy or normal condition. In the effort to understand the meaning of its use in the Moynihan *Report*, one should bear in mind that the therapeutic ideal was every bit as popular in the postwar era as Goffman's construct of the total institution. As a case in point, consider the Warren Commission's effort to comprehend the motives of Kennedy assassin Lee Harvey Oswald. While the commission downplayed Oswald's Marxism, it underscored matters of character and personality, concluding that Oswald was "profoundly alienated from the world in which he lived" and that his "life was characterized by isolation, frustration, and failure."[14] A dead Oswald was unavailable for psychoanalysis, but no such obstacle confronted the psychiatrist in Alfred Hitchcock's *Psycho* (1960). His lengthy, film-ending exegesis presents the murderer, Norman Bates, as the product of an absent father and an overprotective mother—an analysis not only therapeutic but eerily evocative of the Moynihan *Report*'s critique of the black family. Moreover, *Psycho* was only the most significant of dozens of films produced in the 1940s, 1950s, and early 1960s that featured problem-solving psychiatrists or protagonists in need of psychiatric care; among the others were *The Snake Pit* (1948), *Home of the Brave* (1949), *The Caine Mutiny* (1954), *The Searchers* (1956), *Pressure Point* (1962), and, at the end of the cycle, *Captain Newman, M.D.* (1963), in which psychiatry and the therapeutic have lost much of their magic.[15]

What does it mean to understand the Moynihan *Report* as engaging in therapeutic discourse and doing so at a moment in the mid–1960s when, in the cinema at least, that discourse had lost most of its explanatory power? The therapeutic is a personal, rather than a political, discourse. For example, to place the problem of juvenile delinquency in a therapeutic frame, as does the film *Rebel without a Cause* (1955), is to suggest that juvenile delinquents have no politics, that their claims to attention are superficial.[16] Similarly, to put the complaints of women in the 1940s and 1950s in a therapeutic frame is to submit that women of the era had no grievance that a good psychiatrist, armed with Freud's *The Interpretation of Dreams* (1900), could not resolve.[17] And to depict Lee Harvey Oswald's assassination of John Kennedy as the product of an alienated failure was to deny Oswald his politics and, ultimately, to deny the nation an explanation that matched the seriousness of the offense.[18]

Is the Moynihan *Report* of the same ilk? Yes and no. No, in the obvious sense that Moynihan seems to be offering us a social problem—the problem of the black family—rather than a personal one, and his call for national action anticipates a national legislative agenda and a political solution. Moynihan is not proposing to put every black family in group therapy or every black male on the couch. Even so, the *Report* bears witness to Moynihan's limited faith in the efficacy of politics. In an effort to explain the relationship between the therapeutic and the political, he writes: "But here is where the true injury has occurred: unless this damage is repaired, all the effort to end discrimination and poverty and injustice will come to little." In short, therapy must precede politics.[19] Moreover, the problems laid bare in the *Report* are the same ones that had been at the center of the therapeutic discourse in its postwar heyday: weak fathers, overly strong mothers, aggressive women, parental irresponsibility, children with personality defects, ineffective families. Indeed, the *Report*'s catalog of black family flaws reads like the screenplay for *Rebel without a Cause*, which a decade earlier had examined the white middle-class family through the lens of the therapeutic.[20]

The *Report* prefigures an important and obviously political moment—Lyndon Johnson's June 4, 1965, Howard University address. As Steven Steinberg has argued, that address, written by Moynihan and Richard Goodwin, replicated the fault lines of the Moynihan *Report*. While announcing the need for "equality as a fact and as a result," the address was replete with damage imagery and echoes of the therapeutic: centuries of oppression, the "breakdown of the Negro family structure," single-parent families, and ability "stunted" by a "hundred unseen forces playing upon the infant, the child, and the man." And like the *Report*, Johnson's speech suggested an appropriate ordering of the therapeutic and political. "Unless we work to strengthen the family," said the president, "all the rest: schools and playgrounds, public assistance and private concern, will never be enough to cut completely the circle of despair and deprivation."[21]

The ambivalent liberalism of the Moynihan *Report* was a doomed effort to respond to and contain the burgeoning aspirations of black Americans, using a therapeutic discourse that was better suited to the quiescent 1950s, when a bit of Freud seemed to be all that was necessary to cure the frustrated housewife, mollify the alienated teenager, or reinvigorate the bored executive. Nonetheless, elements of the *Report*'s analysis continued to have influence in the social sciences. Despite attacks, damage imagery remained influential, even among racial liberals, who held racism responsible for black pathology. By 1970 scholarly claims of innate black inferiority added another dimension to the argument. Although Moynihan's use of the term "pathology" was widely criticized in the aftermath of the *Report*, its essence seems to have lived on in Edward Banfield's *The Unheavenly City* (1970) and other offerings in the culture-of-poverty school. Indeed, Michael Katz describes the term "culture of poverty" as a "euphemism for the pathology of the undeserving poor" and argues that poverty research has generally "focused more on pathology than on politics."[22] After 1980, conservatives would claim that the welfare state had damaged blacks. Daryl Scott concluded in 1997 that the "emerging consensus within the social science community" was that "many blacks are damaged."[23]

Attacking the Welfare State: The Rise of Economics

In the 1970s and 1980s the psychology that underpinned the Moynihan *Report* yielded to a new emphasis on economics. Two of the major works on social welfare of the post-Moynihan era, Charles Murray's *Losing Ground* (1982) and Martin Anderson's *Welfare* (1978), eschew Moynihan's "pathology" analysis. In describing the hypothetical world of his now-famous inventions, Harold and Phyllis, political scientist Murray insists that "there is no need to invoke the spectres of cultural pathologies or inferior upbringing."[24] Why? Because Harold and Phyllis—exemplars of the urban poor—make "rational" decisions about their lives based on the "same general calculus" of economic and status incentives that everyone else uses, modified by the need of the poor to maximize short-term gains. Applying that rational calculus, in 1960 Harold and Phyllis live together (or get married), Phyllis stays off welfare, and Harold finds a job; applying the same calculus in 1970, after the "reform" era of the Great Society creates a different set of rewards and penalties, Harold and Phyllis do not get married, Phyllis goes on welfare, and Harold stays home. It is a provocative argument, and to make it Murray must reject "pathology"; if Harold and Phyllis have been damaged, they cannot act rationally. In short, Murray rejects pathology not because the behavior of Harold and Phyllis can be explained without it but because their behavior cannot be explained with it.[25]

Similarly, Murray's requirement that his actors respond rationally to incentives meant he must also reject—in this case, simply ignore—Moynihan's model of organizational power and personal passivity, drawn from Elkins's *Slavery*. Yet other writers and scholars, perhaps less wedded to economic instrumentalism, continued to find it useful. Echoes of the Elkins thesis can be found in the culture-of-poverty argument and in its successor concept, the "underclass," heralded by *Time* magazine in 1977 and fully developed in journalist Ken Auletta's *The Underclass* (1982).[26] Another model of black behavior, featuring resilience and resistance rather than defeat and acquiescence, surfaced among liberal social scientists, including historians Herbert Gutman (*The Black Family in Slavery and Freedom* [1976]) and John Blassingame (*The Slave Community* [1972]).[27]

Murray had another reason to reject Moynihan's use of the Elkins slavery/concentration camp comparison or, for that matter, anything similar. Murray argued that an ideological sea change had taken place in 1964. Before that date, Americans of all races and classes believed that the American system was "benign and self-correcting"; after that date, they assumed that "if something was wrong, the system was to blame." *The system was to blame*. It was that flawed doctrine, Murray argues, that made possible the destructive social welfare measures of the next decade, including food stamps, welfare for working people, and additions to public housing.[28] Although Moynihan would be vilified for implying that blacks were somehow responsible for their tattered families ("blaming the victim," in the parlance of the day), the *Report* strongly implied that history had conspired to make black success impossible—in other words, that the system was to blame.[29] Murray could accept that argument, but only insofar as it was applied to the "system" after 1965: the system of the American welfare state.

It should also be noted that *Losing Ground* is without significant historical content. The thesis—that the social welfare programs of the 1960s created the welfare "problem" and harmed the black community—dictated rigorous exclusion of the sort of historical speculation on which Moynihan had grounded his *Report*: centuries of abuse, the trauma of segregation, the ordeal of migration, decades of unemployment. For Murray's argument to work, his blacks had to arrive at 1965 in pretty good shape. In addition, if the federal government were to be understood as the one and only culprit in the welfare melodrama, a way had to be found to rehabilitate—quickly, so that no one would notice—the troubled, de-industrializing economy of the 1970s. Murray accomplished both tasks handily, ignoring slavery, and most of the rest of the African American past, and making a quick-and-dirty case that the 1970s was a decade of considerable economic growth.[30] Murray must have been almost as uncomfortable with Moynihan's thesis, that the black family was in disarray, for again, the timing was wrong; Moynihan's family disintegration was already well under way when Johnson's "reform era" began. As a result, Moynihan gets a mere mention in Murray's chapter, "The Family," and the *Report*'s claims are ever-so-gently dismissed. A dozen lines after introducing the *Report*, Murray is into the "demographic wonder" of black family decline that began, he says, in 1968.[31]

Although colored by its author's deep involvement in Nixon administration welfare policy development in the late 1960s and early 1970s, Anderson's *Welfare* shares several important insights with Murray's book. Like Murray, Anderson downplays the negative economic impact of the economy of the 1970s. Drawing on a wide variety of research, including studies by Edgar Browning, Edward Banfield, Roger Freeman, Sar Levitan, Morton Paglin, and

Alice Rivlin's Congressional Budget Office, Anderson argues that, by 1974, a strong economy and heavy spending on welfare and transfer programs had virtually eliminated poverty.[32] Although Anderson ostensibly credits economic growth with having a hand in this result, it is welfare, and not a general affluence, that centers his claim rhetorically. Welfare, he writes, is a "smashing, total success," a "brilliant success."[33] And it is welfare, and not economic growth, that shoulders the blame in his scenario.[34] This claim not only removes the macroeconomy from critical scrutiny and eliminates poverty as a potential source of social concern, but it sets up the next argument: that the victory in the War on Poverty had "costly social side effects," notably "the almost complete destruction of work incentives for the poor on welfare."[35]

Moynihan's concern with blacks' work values had been minimal, and it was centered on the role of high rates of unemployment in alienating black men and discouraging them from participating in the labor force.[36] Anderson's interest, like Murray's, was of a different sort, and it was based on different assumptions. While Moynihan had assumed that the poor could be psychologically damaged by difficult and desperate conditions experienced over a long period of time, Murray and Anderson assumed that the poor could be instantly affected by economic incentives and disincentives. Both Murray and Anderson make substantial use of the ten-year New Jersey Negative Income Tax experiment, carried out by the Office of Economic Opportunity, and both use the results of that experiment (as well as other studies) to conclude that money payments can dramatically affect the work effort of recipients.[37]

Murray's and Anderson's use of this data reflects a different, and perhaps more optimistic, view of human possibilities than that held by Moynihan. Without expressly offering his own view of the nature of human beings, Anderson tenders and takes issue with Eric Fromm's conviction that work, rather than laziness, is the natural human proclivity.[38] The intense concern with work values and habits may also reflect the close ties that both Anderson and Murray had to the work of libertarian novelist Ayn Rand, whose thought is replete with reflections on the virtue of independence and the evil of dependence.[39] It is also likely that concerns about work reflect doubts about the continued viability of the work ethic in the low-wage, postindustrial service and information economy of the late twentieth century.[40]

Most important, Anderson and Murray celebrate the marketplace; indeed, both acknowledge the importance of free-market economist Adam Smith to their perspectives.[41] In that marketplace perspective, people—the poor, welfare recipients, average Americans, Harold and Phyllis—are never far from being factors of production, whose attitudes toward work, marriage, and family can be shaped, and rather easily shaped, by a few dollars in welfare payments, food stamps, or guaranteed health insurance. Government is harmful not simply because it is the hand of the state but because people are so vulnerable to it.

Gender roles were central to the social analysis of the Moynihan *Report*—so much so, in fact, that one is tempted to conclude that its author was more concerned with patriarchy than with race. On the one hand, Moynihan's investment in patriarchy might be understood as instrumental: a reasonable strategy for black progress. On the other hand, his obsession could be conceptualized as a perverse and misguided effort to apply the logic of the 1940s—of Philip Wylie's *Generation of Vipers* (1942), of Mickey Spillane's *I, the Jury* (1947), or of Marynia F. Farnham's *Modern Woman: The Lost Sex* (1947)—to an era that had already been introduced to Betty Friedan and was about to encounter Gloria Steinem.[42]

In Anderson's *Welfare*, the resistance to working women that shaped the Moynihan *Report* is notably absent, arguably because the emergent postindustrial economy of the intervening years, not to mention feminism, had made that resistance economically and socially untenable. Even so, Anderson's major concern is still with men—now, the anticipated impact of a guaranteed income on the work habits of low-income men. Welfare mothers come within his field of vision but only insofar as their willingness to enter the workplace might be negated by welfare payments. The fact of their motherhood—that these women have children to be raised—does not enter into Anderson's calculations, nor does the possibility that the society might have an interest in paying mothers to stay out of the workplace and raise them. Instead, Anderson implies that work is a sacred and universal social obligation, that everyone physically capable of working should work.[43] "Our welfare programs," Anderson writes, "should be guided by the simple principle that a person gets welfare only if he or she qualifies for it by the fact of being incapable of self-support."[44]

Nonetheless, there is evidence that patriarchy remains a goal of social welfare policy. According to Gwendolyn Mink, the history of welfare reform is the history of efforts to ensure that mothers are dependent on men, not government. To that end, welfare—a system that paid women for raising children and hence allowed them a measure of equality in the family and the marketplace—was gradually redefined as abuse and, in the Personal Responsibility and Work Opportunity Act (1996), all but eliminated. Lacking the assistance of the state, single mothers had to choose between working for wages outside the home or establishing (or reestablishing) a relationship of marital dependency with the child's biological father—the latter, according to Mink, the solution preferred by most reformers. Mink argues that a bias toward patriarchy has influenced social welfare policy since at least 1950, when the first mention of the biological father appears in the statutes. She notes that masculine bias was present in Johnson's Great Society programs, which looked to the employment of fathers as the solution to poverty; in Nixon's proposed Family Assistance Plan (1969), as well as in a Nixon-era change in the welfare laws that allowed the government to require mothers to establish paternity as a condition of receiving payment; and in Reagan's Family Support Act (1988).[45]

The Collapse of Liberalism

In the decade after 1985, liberal, conservative, and libertarian ideas about welfare converged, forging a consensus. Although the consensus did not encompass all those in the social science and policy communities, it was of sufficient breadth to link critical elements of the social science community and to produce dramatic political change—the Personal Responsibility and Work Opportunity Act—under a Democratic president and a Republican Congress. The consensus consisted of a set of interrelated and overlapping ideas and beliefs. Some of these beliefs were general yet pervasive: the centrality of work; the importance of the family; the need for people—the poor, welfare recipients—to be "responsible"; the sense that people could be rather easily changed—damaged or improved—by economic incentives. Others were beliefs about the welfare system: that welfare was the problem rather than a solution; that the provision of welfare encouraged recipients not to work and discouraged the development of two-parent families; that welfare encouraged dependency and irresponsibility; that welfare had caused, or at least contributed to, the making of a distressed, even pathological, black inner-city ghetto.

This consensus might be understood and explained in any number of ways, some of them reflected in the books and documents previously discussed: as an effort, spearheaded by conservatives, to avoid or deny the failures of postindustrial capitalism, revealed in the 1970s; as one product of a general revulsion against the interventionist state, generated by the requirements of competitive global capitalism; as a rationalization of patriarchy in an era of rapidly changing gender roles; as a consequence of the efflorescence of economics within the social sciences; as an attack on the excesses, whether perceived or real, of the 1960s; as an effort, whether appropriate or misguided, to deal with very real changes in late-twentieth-century family structures and in the workplace; and, as I argue below, as a product of a wavering, ambivalent, and troubled liberalism characteristic of the social scientists who were at the center of the Clinton administration's welfare policy.[46]

My focus is on David T. Ellwood, the Harvard School of Government economist who, as assistant secretary for planning and evaluation in Health and Human Services (HHS) in the Clinton administration, cochaired the president's working group on welfare reform with Bruce Reed, a White House adviser, and Mary Jo Bane, a former Harvard colleague who came to HHS as assistant secretary for children and families after serving as commissioner of the New York State Department of Social Services.[47] Clinton administration officials brought Ellwood and Bane on board in part because of their influential early–1980s' collaborative research on welfare and family formation; indeed, Clinton mentioned Ellwood's work during the 1992 campaign.[48] Although Ellwood resigned in 1995 and Bane in 1996 because of differences with the administration over welfare reform legislation, both helped draft the original Clinton welfare plan.[49]

During their earlier academic collaboration, Ellwood and Bane had examined the impact of welfare on family composition, dependency, divorce, and related matters. Their conclusions seemed less than provocative: that very generous AFDC levels encouraged single mothers to live independently; that welfare caused moderate increases in divorce and separation rates; but also that AFDC had little to do with out-of-wedlock births; was not responsible for creation of a dependent, inner-city "underclass"; and, in general, helped protect families and children.[50] Despite the commonplace content of these conclusions, it is widely believed that the Ellwood/Bane research was crucial to the developing liberal-conservative ideological consensus over welfare, first, because it acknowledged that welfare did have some social impact, a conclusion congenial and useful to conservatives, especially because of its origins in the liberal Harvard establishment, and second, because the problem it engaged—the impact of welfare on social and economic formations—bought into conservative assumptions about the role of the welfare state and, concomitantly, did not examine the impact of the postindustrial economy on poor families.[51]

One can imagine how and why conservatives would appropriate the Ellwood/Bane message for their own purposes. But why did these scholars frame problems in ways that invited appropriation? And how, given the unexceptional quality of their conclusions, did they come to favor radical welfare reform? Some insight into these questions can be gleaned from Ellwood's *Poor Support: Poverty in the American Family* (1988). In its double entendre, the title describes Ellwood's ambivalent posture on welfare: it is at once a system that supports the poor and does so poorly. In the preface, Ellwood reveals that the second meaning in the title, and his critical posture toward welfare, developed in the mid–1980s, when the "public" rejected his position. "The message didn't sell very well," he writes. "People hated welfare no matter what the evidence. It wasn't just conservatives; liberals also expressed deep mistrust of the system, and the recipients themselves despised it."[52] It is an extraordinary admission: faced with opinions that differed from his research, Ellwood modified his views and became a critic of welfare.

This change of perspective required some sort of ideological repositioning, and Ellwood offers us glimpses of that process. "Welfare," he explains, "brings some of our most precious values—involving autonomy, responsibility, work, family, community, and compassion—into conflict." And when values come into conflict—when compassion conflicts with autonomy, responsibility, and work—which ones are to have priority? Here is the conflict: "We want to help those who are not making it but, in so doing, we seem to cheapen the efforts of those who are struggling hard just to get by. . . . We want to help people who are not able to help themselves but then we worry that people will not bother to help themselves." As formulated here, the conflict might be resolved either way, in favor of compassion or autonomy. But only pages later, in a list of four "basic American values," compassion is not one of them. Ellwood has made his choice.[53]

As his selection of examples and rhetoric suggests, work ranks high on Ellwood's scale of values. At bottom, compassion threatens to undermine the work ethic. Reprising arguments made by Murray, Anderson, and Lawrence Mead, Ellwood claims that welfare for healthy persons "inevitably reduces the incentive to work" and that welfare for the working poor "conflicts with work; it does not reinforce it."[54] In an essay published a year earlier, Ellwood's commitment to work overrode even the original justification for welfare: assistance to single mothers with children. "Both feminists and psychologists argue," Ellwood asserted, "that some work can be a very valuable thing for both woman and child." In that essay and in *Poor Support*, Ellwood sought to determine how much work society could reasonably expect from an individual or a family unit and, no less important, whether that amount of work at the existing minimum wage would be sufficient to eliminate poverty. Concluding that reasonable work at the minimum wage meant only continued poverty, Ellwood fell back on economic growth and tapped his liberalism for other solutions, including raising the minimum wage and expanding the earned income tax credit.[55] But this did not prevent him from endorsing time-limited welfare, what one scholar has called "the ultimate work-enforcement mechanism."[56]

Ellwood's new commitments—to work, responsibility, and family—were obviously incomplete. Indeed, *Poor Support* might best be understood as a dialogue between the two sides of Ellwood's fragmented political self: on one side, a decaying liberalism; on the other side, an emergent and inchoate libertarianism. This internal dialogue is everywhere in the book. Take the issue of the impact of welfare on family formation—that is, on single-parent families, on divorce, on male responsibility for children, and on women's participation in the labor force. As a liberal social scientist, Ellwood believes/knows that welfare is not a critical factor in any of this. He argues that the link between welfare and single-parent families is "wildly exaggerated," that "the welfare system has had little effect on the structure of families," and that the shape of the late-twentieth-century family is largely a product of macroeconomic policy (a subject, as we have seen, nearly ignored by Murray and Anderson).[57] The argument seems solid enough, but it will not hold. Again, Ellwood's frustration with public opinion, and perhaps with his own loss of authority, looms large. "Most people simply did not believe these studies," he explains. Searching for a middle ground, he admits that the studies point in the "direction" of affecting the family, acknowledges that welfare "could" produce some single-parent families, and finds the liberal argument, that welfare "does more good than harm," "both unappealing and unconvincing."[58]

Likewise, Ellwood's presentation of the role of macroeconomic policy is conflicted. In an essay published in 1987, Ellwood had argued that macroeconomic factors—the stagnant economy of the 1970s and high rates of unemployment—were important ingredients in family formation. Similarly, in *Poor Support* Elwood insists that the family is an inadequate linchpin for the eradication of poverty and that only major changes in macroeconomic policy were likely to produce substantial improvements in harmful social patterns.[59] Yet Ellwood's macroeconomic solutions are far removed from the liberal reform enthusiasm of the Great Society. Reciting a list of the problems usually mentioned by liberals ("a shortage of real jobs—jobs with a future," "people . . . trapped by limited opportunities," "poor education," "discrimination"), Ellwood concludes that the list does not explain why people fail, and he implies that it offers little insight into how people might succeed. As an alternative to liberal scenarios, Ellwood suggests a recast system of supports that would ensure "that everyone who exercises reasonable responsibility can make it without welfare" and a return to early–1960s' style (that is, a rising tide lifts all boats) economic growth, labeled here in Reaganesque terms. "For a large segment of the poor," Ellwood concludes, "trickle-down really works."[60]

Caught between his liberal beliefs about welfare and a public that does not find them credible, Ellwood comes to advocate changes in policy for their anticipated impact on public opinion. "In making changes in the welfare system," Ellwood writes in one of *Poor Support*'s most tortured passages,

> one certainly ought to consider the incentives they create for the formation of single-parent families. That is the essence of the welfare-

> family structure conundrum. Such incentives are important, not only for their real effects, but for their effects on the perceptions of the public at large. This book recommends major changes in the welfare system that might even reduce the number of single-parent homes. But let us not be fooled into thinking that marginal changes in welfare are likely to have much effect on the powerful forces that are influencing the formation of single-parent families.[61]

In short, Ellwood argues that welfare must be reformed not because the changes will affect families or children in any important way but in order to change the perception that the welfare system had caused damage.

The central themes of *Poor Support*—the importance of perception, a muddled liberalism—come together in a chapter on ghetto poverty. On the one hand, Ellwood finds the concept "ghetto poverty" imperfect. He denies that the ghetto is truly representative of American poverty. And he suggests that the image of the black ghetto as a place of aberrant behavior and damaged values, captured in Auletta's term "underclass," was a media creation that distorted reality.[62] On the other hand, Ellwood finds the idea irresistible and useful. Notwithstanding his critique of the media, Ellwood offers the reader excerpts from Leon Dash's sensational *Washington Post* series on the inner city, using these stories to demonstrate that the ghetto sets limits on thought and behavior. Relying on the social psychological framework of Thomas Kane, he depicts the ghetto as a space of uncontrollable contingency. In the 1960s, he argues, Kenneth Clark, Herbert Gans, Oscar Lewis, and Moynihan (in his *Report*) were on the right track; they had uncovered the problem of social disorganization in the ghetto, and they were optimistic that solutions could be found to it. Unfortunately, Ellwood adds, liberal criticism of the Moynihan *Report* for "blaming the victim" derailed inquiries into ghetto behavior until the 1980s, when conservatives were once again willing to talk about "bad values" among the poor. And that is where Ellwood comes in, sure now that the time has come for liberals to join conservatives in examining the flawed "behavior, attitudes, and expectations of ghetto residents." Sounding very much like Charles Murray, Ellwood concludes that it is time to stop "blaming only the system" and to find ways to "reward responsibility and initiative."[63]

Conclusion

What do these texts reveal about the social sciences and welfare in the late twentieth century? They tell us that damage imagery and the therapeutic perspective, so potent in the Moynihan *Report*, linger on in the 1990s as its legacy, even while serving a different, and more conservative, constituency. They reveal how the libertarian right, represented here in texts by Anderson and Murray, was able to shift the terms of the welfare debate, abandoning Moynihan's racial, historical, and masculinist frameworks for a nonracial, presentist, and more gender-neutral approach that presented welfare recipients as actors in a marketplace, responsive to even short-term economic stimuli.

There was renewed optimism in this marketplace perspective. Where Moynihan had depicted blacks as victimized and damaged by decades of discrimination and unemployment, Anderson and Murray ignored that heritage and neglected the profound dislocations that characterized the post–1970, postindustrial American economy. And where Moynihan had found a deeply rooted social problem—the pathological black family—that threatened to defeat the liberal agenda, Anderson and Murray found people wrongly and perversely motivated by a welfare system that could be fixed. Although Anderson and Murray agreed that welfare was the problem, they differed somewhat on what it was that welfare did. They shared the conviction that welfare corrupted the work ethic; Murray added the important argument that welfare damaged the family.

These two arguments—that welfare injured families and made people irresponsible—proved too much for 1980s' liberalism. The first of these arguments was misleading if not fallacious, yet it proved remarkably resilient nonetheless. We have observed a weary David Ellwood, our representative of liberalism, in a futile effort to convince the public—and in the end, himself—of welfare's minimal impact on family arrangements. The second argument, about welfare and work, proved no less powerful. Moynihan's anxieties about a matriarchy caused by working women now seemed beside the point. So, too, did the concerns of Progressive Era and New Deal reformers who had shaped the state and federal welfare systems for half a century: that mothers needed and deserved state support to remain out of the workplace and to provide care for their dependent children. In the low-wage, full-employment economy that characterized most of the 1980s and 1990s—an economy in which most adults *had* to work and most adults *could* work—work became a natural, even irresistible solution to social problems, even for liberals. Unable to defend the liberal agenda or to explain why welfare had once been deemed essential, Ellwood—social science's equivalent of an exhausted Michael Dukakis, laboring to uphold the liberal heritage in the 1988 presidential debates—gave up.

Notes

1. U.S. Department of Labor, Office of Policy Planning and Research, *The Negro Family: The Case for National Action*, March 1965 (N.p., n.d.), 1–3 (hereafter cited as Moynihan *Report*).

2. Ibid., page preceding table of contents, 6, 19.
3. Ibid., 16, 47.
4. John W. Blassingame, *The Slave Community: Plantation Life in the Antebellum South* (1972; rev. and enlarged ed., New York: Oxford University Press, 1979), 285–331.
5. Christopher Lasch, *The Minimal Self: Psychic Survival in Troubled Times* (New York: W. W. Norton, 1984), 112.
6. Ibid., 113.
7. Erving Goffman, *Asylums: Essays on the Social Situation of Mental Patients and Other Inmates*, Anchor Books ed. (Garden City, N.Y.: Doubleday, 1961).
8. Moynihan *Report*, 47.
9. Ibid., 16. On the dozens, see Henry Louis Gates Jr., "2 Live Crew, Decoded," *New York Times*, June 19, 1990.
10. Moynihan *Report*, 16, 19, 20, 25, 29.
11. Ibid., 39, 34–38.
12. Adapted from William Graebner, *The Age of Doubt: American Thought and Culture in the 1940s* (Boston: Twayne, 1991), 92.
13. Daryl Michael Scott, *Contempt and Pity: Social Policy and the Image of the Damaged Black Psyche, 1880–1996* (Chapel Hill: University of North Carolina Press, 1997), xii, 129, 135–136.
14. *The Warren Report: Report of the President's Commission on the Assassination of President John F. Kennedy* (N.p.: Associated Press, n.d.), 160. Historian Richard Hofstadter used the term "political pathology" to describe nonrational styles of political discourse—and hence to establish those discourses as outside the consensus—in *The Paranoid Style in American Politics* (New York: Alfred A. Knopf, 1965). Two decades later, Moynihan, then a senator from New York, applied Hofstader's framework to the right wing of the Republican Party. Daniel Patrick Moynihan, "The Paranoid Style in American Politics Revisited," *Public Interest* 81 (1985): 107–127. See also Mark Fenster, *Conspiracy Theories: Secrecy and Power in American Culture* (Minneapolis: University of Minnesota Press, 1999), 3–21.
15. Krin Gabbard and Glen O. Gabbard, *Psychiatry and the Cinema* (Chicago: University of Chicago Press, 1987); Janet Walker, *Couching Resistance: Women, Film, and Psychoanalytic Psychiatry* (Minneapolis: University of Minnesota Press, 1993). See also Elizabeth Lunbeck, *The Psychiatric Persuasion: Knowledge, Gender, and Power in Modern America* (Princeton, N.J.: Princeton University Press, 1994).
16. The point is nicely made in the "Officer Krupke" number in *West Side Story* (film version, 1961). On how youth gangs were understood in the framework of psychopathology, see Eric C. Schneider, *Vampires, Dragons, and Egyptian Kings: Youth Gangs in Postwar New York* (Princeton, N.J.: Princeton University Press, 1999), 12–18.
17. Sigmund Freud, *The Interpretation of Dreams* (1900; London: Hogarth, 1953).
18. Max Holland, "After Thirty Years: Making Sense of the Assassination," *Reviews in American History* 22 (1994): 191–209.
19. Moynihan *Report*, 5.
20. James Gilbert, *A Cycle of Outrage: America's Reaction to the Juvenile Delinquent in the 1950s* (New York: Oxford University Press, 1986), 185–188.
21. The quotations are from Stephen Steinberg, "The Liberal Retreat from Race," *New Politics* 5 (summer 1994), Internet version at www.wilpaterson.edu/~newpol/issue17/steinb17.htm, 5–6. See also Scott, *Contempt and Pity*, 151.
22. Michael B. Katz, *Improving Poor People: The Welfare State, the "Underclass," and*

Urban Schools as History (Princeton, N.J.: Princeton University Press, 1995), 70, 71.

23. Scott, *Contempt and Pity*, 182, 181, 188, 198.
24. Charles Murray, *Losing Ground: American Social Policy, 1950–1980* (New York: Basic Books, 1984), 162; Martin Anderson, *Welfare: The Political Economy of Welfare Reform in the United States* (Stanford, Calif.: Hoover Institution Press, 1978).
25. Murray, *Losing Ground*, 154–155. There is a hint of the pathology argument in Murray's discussion of the way in which the national government and the National Welfare Rights Organization, by removing the welfare stigma and denying the importance of personal responsibility, damaged black self-confidence (178–191).
26. Katz, *Improving Poor People*, 63–65.
27. Scott, *Contempt and Pity*, 162; Herbert Gutman, *The Black Family in Slavery and Freedom* (New York: Vintage Press, 1976); Blassingame, *The Slave Community*.
28. Murray, *Losing Ground*, 44, 46.
29. The classic study, and perhaps the origin of the phrase, is William Ryan, *Blaming the Victim*, rev. ed. (New York: Viking, 1976).
30. Murray, *Losing Ground*, 58–59.
31. Ibid., 129–130, and ch. 9, "The Family."
32. Anderson, *Welfare*, 15, 19–25, 37. Rivlin's CBO would play an increasingly important role in providing information for welfare policy-makers, especially those on the right. In the 1980s Rivlin would be a member of the Working Seminar on Family and American Welfare Policy, chaired by Michael Novak, director of social and political studies for the American Enterprise Institute. See Novak et al., *The New Consensus on Family and Welfare: A Community of Self-Reliance* (Washington, D.C.: American Enterprise Institute for Public Policy Research, 1987).
33. Anderson, *Welfare*, 38, 39.
34. Ibid.
35. Ibid., 43.
36. Moynihan *Report*, 43, 44.
37. Murray, *Losing Ground*, 148–153; Anderson, *Welfare*, 102–105, 119.
38. Anderson, *Welfare*, 90–91.
39. See Ayn Rand, *The Fountainhead* (1943; New York: Signet, 1971), 686–689.
40. Charles Murray, "The Local Angle: Giving Meaning to Freedom," *Reason Magazine*, February 2, 1999, Internet version at www.reasonmag.com/murray25speech.html.
41. Anderson, *Welfare*, vii; Charles Murray, *What It Means to Be a Libertarian: A Personal Interpretation* (New York: Broadway Books, 1997), xiii.
42. Scott, *Contempt and Pity*, 153–154; Graebner, *Age of Doubt*, 106–107.
43. Anderson, *Welfare*, 98–100.
44. Ibid., 163.
45. Gwendolyn Mink, *Welfare's End* (Ithaca, N.Y.: Cornell University Press, 1998), 35, 23, 40, 36, 42–43.
46. For evidence that the 1960s were the target of the politics of consensus, see Novak et al., *The New Consensus*, where the value of "self-control" is contrasted with an earlier era in which "self-control and impulse restraint were debunked as 'square'" and "'self-expression' was portrayed as a 'higher form of consciousness'" (14).
47. David T. Ellwood, "Welfare Reform as I Knew It: When Bad Things Happen to Good Policies," *American Prospect* 26 (May-June 1996): 22–29. The citation is to the Internet version at http://www.epn.org/prospect/26/26ellw.html, 1. See also Barbara Vobejda and Judith Havemann, "2 HHS Officials Quit Over Welfare Changes," *Washington Post*, September 12, 1996, Internet version at washington post.com/wp-srv/national/longterm/welfare/quit.htm; and Mary Jo Bane biography, at http://www. excelgov.org/chap_hc_n2.htm.

48. Sanford F. Schram, *Words of Welfare: The Poverty of Social Science and the Science of Poverty* (Minneapolis: University of Minnesota Press, 1995), x; Ellwood, "Welfare Reform," 1.
49. Kelley Rouse, "Living with Conscience: The Edleman Resignation," *Shore Journal*, September 22, 1996, Internet version at http://www.shorejournal.com/9609/kjr0922a.html, 1–2. See also E. J. Dionne Jr., "Resigning on Principle," *Washington Post*, September 17, 1996, Internet version at 206.132.25.71/wp-srv/politics/special/welfare/stories/op091796.htm
50. Michael Morris and John B. Williamson, *Poverty and Public Policy: An Analysis of Federal Intervention Efforts* (Westport, Conn.: Greenwood Press, 1986), 59; Schram, *Words of Welfare*, xi, 55–57; David T. Ellwood, *Poor Support: Poverty in the American Family* (New York: Basic Books, 1988), ix.
51. Schram, *Words of Welfare*, 10, 131, 219 n.47, 13. For examples of the appropriation of liberal research to conservative ends, see Novak et al., *The New Consensus*, 9 (on Mary Jo Bane); and Lowell Gallaway and Richard Vedder, "The Impact of the Welfare State on the American Family," prepared for the Joint Economic Committee, 1996, Internet version at http://www.house.gov/jec/welstate/vg–5/vg–5.htm, n.18, citing Ellwood and Bane's 1982 research on AFDC and family structures.
52. Ellwood, *Poor Support*, ix–x.
53. Ibid., 6, 16–17.
54. Ibid., 237, 127, 104.
55. David T. Ellwood, *Divide and Conquer: Responsible Security for America's Poor*, Occasional Paper No. 1, Ford Foundation Project on Social Welfare and the American Future (New York: Ford Foundation, 1987), 37, 3–4, 19; Ellwood, *Poor Support*, 87–88, 109–110, 14–15. On Ellwood's view of work, see also Ellwood, "Welfare Reform as I Knew It," 2 ("From the moment someone walks through the door [of a welfare office], every signal ought to be that work is the ultimate goal and expectation").
56. Joel F. Handler, *The Poverty of Welfare Reform* (New Haven, Conn.: Yale University Press, 1995), 5–6, 149–150.
57. Ellwood, *Poor Support*, 42, 22, 73.
58. Ibid., 22–23, 26.
59. Ellwood, *Divide and Conquer*, 11–12, 15; Ellwood, *Poor Support*, 72–73.
60. Ellwood, *Poor Support*, 8, 11, 73, 96–98. Martin Anderson traces the term "trickle down" to a single reference by Reagan appointee David Stockman, reported in the *Washington Post*. Otherwise, he argues, no one in the Reagan administration, including the president, ever used the term. See Martin Anderson, "Social Welfare Policy: The Objectives of the Reagan Administration" (Stanford, Calif.: Hoover Institution Press, n.d.).
61. Ellwood, *Poor Support*, 77–78.
62. Ibid., 194, 189, 195.
63. Ibid., *Poor Support*, 212–214, 196–197, 199, 225–226. The Working Seminar on Family and American Welfare Policy, a conservative group that included Charles Murray and Lawrence Mead, was attracted to Auletta's work because it supported the idea of a black underclass suffering from behavioral problems. See Novak et al., *The New Consensus*, 11, 36, 219.

Afterword

ANY FINAL ASSESSMENT of the question of the social sciences in Washington, so to speak, can hardly be entertained now. Such would be premature, to say the least. The contributors in this volume have only laid out a possible sketch. Yet several things do appear to be clear, even at this preliminary stage.

It is manifestly obvious that the world has changed since the middle decades of the twentieth century. Now, in the dawn of the twenty-first century, we can recognize that the world of American politics has changed, and massively so, since the 1940s and 1950s—when the social sciences seemed to enjoy a rising influence on national public policy and within the contemporary national, state, and local political systems. It was in the 1830s and 1840s that what political historians have called "the second party system" took shape in American national life, that party system in which parties were coalitions of various interests—of particular groups in political economy, to be sure, such as national versus state bankers, of high- and low-tariff men, of those for states rights and those for a Whiggish national political economy, and the like. From then on, Americans had several ways in which to participate in politics and government and thus to use an instrumentalist approach to politics—how can a person as an individual or as a member of an interest group use the political system to obtain public goods for oneself or one's interest group? Thus the traditional method of soliciting aid from the gentry or the government through individual petitions remained in place, but it was joined by a new method of participation in national politics, not merely supporting an ideological leader, as had happened in the Federalist-Jeffersonian era, but supporting a party of leaders (and followers) who represented and constituted a coalition of various interest groups. It remains an interesting, indeed tantalizing, point that if the 1830s and 1840s witnessed the invention of interest-group political parties, they also saw the simultaneous emergence of the notion of the group as a building block of society and of theories about society, along with the invention of statistics, or group dynamics and characteristics, at the same time. Regrettably, we cannot pursue such historical coincidences, if coincidences they were, in the limited space available here.

The larger point is that in the last five or six decades our national political system in general and our national political parties in particular have changed from being coalitions of various institutional and group interests to being increasingly ideological "single-issue" constituencies, especially with regard to cultural and social issues—education, welfare, the environment, race relations, feminism, the so-called culture wars over science and education, and the like. And the political parties have in another sense become the personal vehicles of politicians who follow entrepreneurial and individualistic careers rather than the corporate and bureaucratic political careers that party men and women pursued for a century or more. Hence the champions of social scientific public policy have often been professional social scientists and their allies versus whatever entrenched interests oppose such innovations, whether in the private sector corporations whose officers and functionaries are involved in public policy or those public bureaucracies whose leaders and minions oversee public policy. Given the individualistic tendencies of American politics and what Zane Miller rightly labels the cultural individualism of national life, it is hardly surprising that so many public policy projects informed by social science, as those whose histories are included in this volume, dissipate into cultural expressions of personal choices. Indeed, in various ways, such seems the common fate of all such operations in the preceding chapters. In Hamilton Cravens's essay, African Americans never really developed a means of becoming an interest group within the political and constitutional system. Michael Bernstein's innovative economists remained individual advisers, without sufficient institutional clout to make peace, not war, a genuine possibility. Harvey Sapolsky's military and civilian bureaucrats create a system of argumentation and presentation within governmental circles that outflanks the traditional interest-group lobbying techniques. Howard Segal and Philip Frana, in equally profound and striking case examples, show how public health becomes private health and "wellness" and how big technology becomes merely more consumer toys (hardly what the great C. P. Snow had in mind). As Hal Rothman argues, there were certain economic limits—of individual self-interest—to the environmental movement. Such programs as Head Start, in the chapter by Nawrotzki, Smith, and Vinovskis, remained more a political symbol than a substantive reform, urban policy disintegrated into a hyperdiversity and extreme cultural individualism that meant ceaseless social fragmentation, and, in Graebner's telling, the jettisoning of welfare meant the return of a Dickensian present and future for the poor of America. In short, the common welfare seems no longer to exist, whether social science has anything to do with such at all; it is cultural individualism *über alles*, and with gusto.

The simple truth is that *public* policy cannot be created for, or maintained for, individuals or individual needs. To have national public policy there must be a sense of a larger whole, a national citizenry that binds all in society together despite wide differences of agreement over many specific interests. Such appears to be lacking in contemporary America. Indeed, such seems to define what is "postmodern" about post–World War II American national culture. Another difficulty is that without a shared national public culture it is very difficult to have a shared sense of truth, objectivity, authority, expertise, and the like, and their corresponding opposites, such as falsity, partisanship, and the like. Hence the roots of the postmodernist crisis of authority, of positive truth, of science and its utility, not to mention its contributions to truth, beauty, and justice. What may have begun as an attempt, at mid-twentieth century, to reform politics as they are and bring objectivity, authority, and truth to national policy may well have dissipated into fragmentation, individuation, and personal expression.

One matter remains clear. The social sciences are with us for some time to come. Whether the postmodernist mood is a temporary tantrum or whether it really does denote a new age, different qualitatively from that which preceded it in multitudinous ways, is still something that may be honestly debated. I happen to believe that we have entered a new and different era and that the old rules of conduct and judgment no longer apply. But that hardly means that the institutions and social practices of that past age no longer exist but more likely that they probably work in ways most of us have still not come to understand and thus to master. Such has been the character of any historical shift in the *Geist* of which I am aware, and I see no reason to think that matters would be different now.

Contributors

MICHAEL A. BERNSTEIN is professor of history and associated faculty member in economics at the University of California, San Diego. His research and teaching focus on the modern economic and political history of the United States. His most recent book is *A Perilous Purpose: Economists and Public Purpose in Twentieth-Century America* (2001).

HAMILTON CRAVENS is professor of history, Iowa State University. His research and teaching interests embrace the history of science, technology, and medicine, and of public policy, in nineteenth- and twentieth-century American culture. His most recent book is *Before Head Start: The Iowa Station and America's Children* (2002).

PHILIP L. FRANA is software history project manager, Charles Babbage Institute, University of Minnesota–Twin Cities. His research and teaching interests are the history of science, medicine, and technology in modern America, especially the history of information technology and bioinformatics, particularly medical informatics. Among his publications are the forthcoming "Hunting the Wumpus: Internet Gopher and the Information Revolution," *IEEE Annals of the History of Computing*.

WILLIAM GRAEBNER is professor of history at the State University of New York, College at Fredonia. His contributions to the history of public policy include *Coal-Mining Safety in the Progressive Period: The Political Economy of Reform* (1976) and *A History of Retirement: The Meaning and Function of an American Institution, 1885–1978* (1980). He is associate editor of *American Studies*.

ZANE L. MILLER is Charles Phelps Taft Professor of History Emeritus, University of Cincinnati. He is a leading teacher and scholar of American urban history, and his most recent book is *Visions of Place: The City, Neighborhoods, and Cincinnati's Clifton, 1850–2000* (2001). Among his many other books is the classic *Box Cox's Cincinnati: Urban Politics in the Progressive Era* (1968).

KRISTEN D. NAWROTZKI is currently completing her Ph.D. at the University of Michigan in American social history. She is a lecturer at the School of Education, Heidelberg, Germany, where her research and teaching focus on the history of education in the United States and Britain.

HAL ROTHMAN is professor of history and chair of the Department of History, University of Nevada–Las Vegas. A prolific and acknowledged expert in American environmental history, his most recent books are the runaway best-sellers, *Neon Metropolis: How Las Vegas Started the Twenty-first Century* (2001) and *Devil's Bargains: Tourism in the Twentieth-Century West* (2000).

HARVEY M. SAPOLSKY is professor of political science at the Massachusetts Institute of Technology. His teaching and research focus on science and technology in the military and their international consequences. Among his many books is *Science and the Navy: The History of the Office of Naval Research* (1990).

HOWARD P. SEGAL is a Bird and Bird Professor of History at the University of Maine–Orono. He teaches and writes about technology in American culture. Among his most recent books is *Technology in America: A Brief History* (1999), of which he is the coauthor, and *Recasting the Machine Age: Henry Ford's Village Industries* (2004).

ANNA MILLS SMITH is currently completing her doctorate in American social history at the University of Michigan. Her research and teaching interests focus on the history of children and youth in America.

MARIS VINOVSKIS is Bentley Professor of History at the University of Michigan. Among his many books are *History and Educational Policymaking* (1999) and *Revitalizing Federal Educational Research and Development: Improving the R&D Centers, Regional Education Laboratories, and the New OERI* (2001).

Index